D0918410

# Drilling Discharges in the Marine Environment

Panel on Assessment of Fates and Effects of Drilling Fluids and Cuttings in the Marine Environment

Marine Board

Commission on Engineering and Technical Systems

National Research Council

NATIONAL ACADEMY PRESS
Washington, D.C. 1983

National Academy Press • 2101 Constitution Avenue, N.W. • Washington, D.C. 20418

NOTICE: The project that is the subject of this report was approved by the Governing Board of the National Research Council, whose members are drawn from the councils of the National Academy of Sciences, the National Academy of Engineering, and the Institute of Medicine. The members of the panel responsible for the report were chosen for their special competences and with regard for appropriate balance.

This report has been reviewed by a group other than the authors according to procedures approved by a Report Review Committee consisting of members of the National Academy of Sciences, the National Academy of Engineering, and the Institute of Medicine.

The National Research Council was established by the National Academy of Sciences in 1916 to associate the broad community of science and technology with the Academy's purposes of furthering knowledge and of advising the federal government. The Council operates in accordance with general policies determined by the Academy under the authority of its congressional charter of 1863, which establishes the Academy as a private, nonprofit, self-governing membership corporation. The Council has become the principal operating agency of both the National Academy of Sciences and the National Academy of Engineering in the conduct of their services to the government, the public, and the scientific and engineering communities. It is administered jointly by both Academies and the Institute of Medicine. The National Academy of Engineering and the Institute of Medicine were established in 1964 and 1970, respectively, under the charter of the National Academy of Sciences.

This report represents work supported by Contract 14-12-0001-29063 between the U.S. Department of the Interior and the National Academy of Sciences.

Library of Congress Catalog Card Number 83-62998
International Standard Book Number 0-309-03431-0



Printed in the United States of America

## Panel on Assessment of Fates and Effects of Drilling Fluids and Cuttings in the Marine Environment

John D. Costlow, Chairman
Duke University Marine Laboratory
Beaufort, North Carolina

Robert C. Ayers, Jr.
Exxon Production Research Company
Houston, Texas

Donald F. Boesch
Louisiana Universities Marine Consortium
Chauvin, Louisiana

Thomas R. Gilbert
Northeastern University
Boston, Massachusetts

James G. Gonders
Cities Service Company
Oklahoma City, Oklahoma

Donald W. Hood
Consultant
Friday Harbor, Washington

Kenneth D. Jenkins*
California State University
Long Beach, California

Jerry M. Neff
Battelle New England Marine Research Laboratory
Duxbury, Massachusetts

James P. Ray
Shell Oil Company
Houston, Texas

Hal Scott
ECO/Interface Evaluations
Orlando, Florida

Judith Spiller*
University of New Hampshire
Durham, New Hampshire

Kenneth R. Tenore
Skidaway Institute of Oceanography
Savannah, Georgia

David C. White
Florida State University
Tallahassee, Florida

Staff

Charles A. Bookman, Senior Staff Officer
William Kirby-Smith, Consultant
Phyllis Johnson, Secretary
Delphine D. Glaze, Word Processer

---

*Appointed in September 1982.

# Preface

## ORIGIN OF THE STUDY

A variety of wastes are generated in drilling oil and gas wells, including drill cuttings and used drilling fluids.[1] The disposal of these wastes is licensed by the U.S. Environmental Protection Agency (EPA) under the National Pollutant Discharge Elimination System (NPDES) (40 CFR 122-125). Permitting such discharges on the outer continental shelf (OCS)[2] has occasioned considerable public debate about how much they may harm the marine environment.

To improve the technical basis for decision making about discharging drilling fluids and cuttings in the marine environment, the Bureau of Land Management[3] turned to the National Research Council for a critical review of the subject. In response, the Assembly of Engineering[4] of the National Research Council convened the Panel on Assessment of Fates and Effects of Drilling Fluids and Cuttings in the Marine Environment under the auspices of the Marine Board. Members of the panel were selected for their experience in

---

[1] Drilling fluid is also called mud or drilling mud, because it often looks like mud. All these terms are commonly used in the oil and gas industry. For consistency, the term drilling fluid is used throughout this report.

[2] The OCS is that portion of the submerged continental margin that is subject to U.S. jurisdiction. For the purpose of this report, the OCS extends from a state's offshore boundary (3 miles offshore except off Texas and west Florida where state boundaries extend 3 leagues--9 nautical miles--offshore) out to the limit of economic exploitation.

[3] In a subsequent federal reorganization, the sponsorship of this study was transferred to the newly created Minerals Management Service.

[4] In a reorganization of the National Research Council in the spring of 1982, the Assembly of Engineering was subsumed by the newly created Commission on Engineering and Technical Systems.

marine biology, marine environmental analysis, toxicological studies of marine animals, chemical oceanography, benthic ecology, the technology and chemistry of drilling fluids, and offshore drilling operations. Consistent with the policies and procedures of the National Research Council, appropriate balance of perspectives was an important consideration in choosing panel members.

## SCOPE OF STUDY

The charge to the panel was to establish a credible technical basis for decisions about discharging drilling fluids and cuttings in the marine environment. The panel proceeded by reviewing and critically appraising the available knowledge concerning the fates and effects of drilling fluids and cuttings on the OCS. It assessed the adequacy and applicability of existing research and the transferability of research results to different sites and hydrodynamic regimes. The panel also considered additional needed research as well as various means to mitigate the potential effects of drilling discharges.

In agreement with its charge the panel focused on discharges made during exploratory and development drilling, as opposed to those made during other phases of OCS operations. It did not consider the fates and effects of the formation waters primarily produced during oil and gas production, nor those of certain specialty drilling fluids used infrequently and in limited quantities during periodic maintenance operations and in preparing wells for production. The panel further restricted its study to water-based drilling fluids, since these are the fluids used in the vast majority of OCS wells and are the only kind of fluids currently permitted to be discharged on the OCS.

Drilling discharges are but one impact on the marine environment from petroleum resource development. In addition to specialty drilling fluids (not covered in this report as explained above), waters from hydrocarbon-bearing formations are discharged during production, and spills or blowouts may occur. Also, petroleum development may compete with other resource uses, such as fishing. It is the combination of factors that contribute to the effects of OCS petroleum development.

The assessment was limited to the OCS. Nevertheless, substantial oil well drilling activity occurs on state lands, and, often similar oceanographic conditions prevail in state waters as on the OCS. Thus, the application of the data and results in the report to state waters (or any other marine environment) is appropriate where the physical conditions that prevail in the state waters are similar to those of the OCS, expecially those reported in OCS field studies. The fates and effects of discharged drilling fluids and cuttings in restricted nearshore waters, such as in estuaries and embayments, were not a subject of this study. The panel's work on the nature of drilling fluids, on considerations in using the available scientific information, and on mitigating measures applies to all marine environments.

## METHOD OF THE STUDY

Available data on the fates and effects of drilling fluids in the marine environment were of paramount importance throughout this study. Panel members initially received four comprehensive, and in some cases critical, reviews of the available literature (Houghton, et al., 1981; Neff, 1981, Petrazzuolo, 1981; Rieser and Spiller, 1981) for a broad overview of current knowledge in the field and attendant technical and public issues. Several panel members had such a long and consistent involvement in the subject that they were also familiar with virtually all the abundant original literature on the subject (a bibliography of this literature is available [IMCO Services, 1982]).

The panel considered all available literature and even current (unpublished) work related to its charge. It found that some aspects of the problem, such as ocean dispersion, were best treated by older well-established literature, while others, such as the toxicity of particular fluids or fluid additives, were discussed only in draft reports that had not yet passed through normal publication procedures. The panel also considered conventional peer-reviewed journal articles and the vast amount of so-called "gray literature," which may have been subjected to various levels of review, but which is limited in circulation and availability. It relied on peer-reviewed literature when such literature was available, but also used the gray literature when its quality could be established. An important aspect of the panel's work was weighing the quality of all these scientific contributions. Another of equal importance, but often more difficult to achieve, was determining the applicability of the research to the problem under consideration. Some of the research reviewed was designed to test a hypothesis; other research had been conducted to satisfy regulatory or other mandates. The panel attempted to accommodate these diverse sources by evaluating the literature on its scientific merit and on its applicability to the objectives of this study.

At the outset the panel solicited public comments on the issues it should address (Federal Register, Oct. 23, 1981); 33 sets of comments were received. With this initial guidance, the panel then conducted its review of the technical literature and summarized this review in a set of discussion papers. About 200 copies were distributed for review, and 46 substantive written reviews were received. An open meeting of the panel provided additional opportunity for interested persons to identify and discuss related issues. Seventy people attended the day-long meeting. The panel then sought additional information to address concerns the public had raised and to complete its assessment.

Thus, the panel's report and its conclusions and recommendations are based on its review of the primary and secondary scientific literature, on the public comments that were received, on additional data and information sought by the panel, and on the professional experience of panel members.

ACKNOWLEDGMENTS

The panel benefited greatly from the interest, material contributions, intellectual challenges, and encouragement of a number of individuals and organizations. In addition to the many reviewers and commenters, the following made especially important contributions to the panel's work: the American Petroleum Institute, Michael Connor (Harvard School of Public Health), Tom Duke (Environmental Protection Agency), William Grant (Woods Hole Oceanographic Institute), Maurice Jones (IMCO Services Division, Halliburton Company), Burt Keenan (National Advisory Committee on Oceans and Atmospheres), Gary Petrazzuolo (Technical Resources), and Robert Spies (Lawrence Livermore Laboratories). The panel also wishes to acknowledge the contributions of the liaison representatives of several government agencies: James Cimato (Minerals Management Service), Joseph Kravitz (National Oceanic and Atmospheric Administration), Douglas Lipka (Environmental Protection Agency), and Edward Tennyson (Minerals Management Service).

REFERENCES

Houghton, J.P., et al. 1981. Fate and effects of drilling fluids and cuttings discharges in lower Cook Inlet, Alaska, and on Georges Bank. Report prepared for Branch of Environmental Studies, Minerals Management Service. Dames & Moore, Inc., Seattle, Wash.

IMCO Services. 1982. Environmental Aspects of Drilling Fluids: A Bibliography. 3d ed. Technical Bulletin. Houston, Tex.: IMCO Services Division, Halliburton Company.

Neff, J.M. 1981. Fate and biological effects of oil well drilling fluids in the marine environment: a literature review. Draft report (15-077) to the U.S. Environmental Protection Agency. Environmental Research Laboratory, Gulf Breeze, Fla.

Petrazzuolo, G. 1981. Preliminary report: an environmental assessment of drilling fluids and cuttings released onto the Outer Continental Shelf. Vol. 1: Technical assessment. Vol. 2: Tables, figures, and Appendix A. Draft report prepared for Industrial Permits Branch, Office of Water Enforcement and Ocean Programs Branch, Office of Water and Waste Management, U.S. Environmental Protection Agency, Washington, D.C.

Rieser, A., and J. Spiller. 1981. Regulating Drilling Effluents on Georges Bank and the Mid-Atlantic Outer Continental Shelf: A Scientific and Legal Analysis. Boston, Mass.: New England States/ New England River Basins Commissions. 130 pp.

# Contents

| | Page |
|---|---|
| Summary, Conclusions, and Recommendations | 1 |
| 1. Introduction | 9 |
| 2. Drilling Discharges | 11 |
| Offshore Oil and Gas Drilling and Development | 11 |
| Characteristics and Functions of Drilling Fluids | 13 |
| Discharges of Drilling Fluids | 14 |
| The Compositions of Discharges | 17 |
| Components of Water-Based Drilling Fluids | 15 |
| The Chemistry of Drilling Fluids | 24 |
| Drilling-Fluid Components as Commercial Products | 25 |
| Trends in Operating Practices: Generic Drilling Fluids | 26 |
| The Regulation of Drilling Fluids | 34 |
| The Mass Loading of Drilling Discharges in Relation to That of Other Inputs to the Marine Environment | 37 |
| References | 44 |
| 3. The Fates of Drilling Discharges | 49 |
| Introduction | 49 |
| Behavior of the Discharge Plume | 51 |
| Fates of Drilling Fluids and Cuttings | 56 |
| References | 69 |
| 4. The Biological Effects of Drilling Discharges | 75 |
| Introduction | 75 |
| The Toxicities of Drilling Fluid Components | 75 |
| The Toxicities of Used Drilling Fluids | 84 |
| Bioavailability | 105 |
| Conclusions | 112 |
| References | 112 |
| 5. Considerations in Using the Information Available on the Fates and Effects of Drilling Discharges | 129 |
| Laboratory Evaluations of Toxicity | 129 |
| Bioaccumulation | 132 |

CONTENTS (continued)

| | Page |
|---|---|
| The Variability of Drilling Fluids | 133 |
| Field Studies of the Fates and Effects of Drilling Fluids | 133 |
| Extrapolation of Results | 137 |
| Long-Term Fates and Effects | 137 |
| Other Information | 139 |
| References | 141 |
| 6. Alternative Operating Practices | 147 |
| Shunting | 147 |
| Dilution Requirements and Limitations on Rates of Discharge | 149 |
| Offloading and Transport for Distant Discharge | 149 |
| Other Transport Techniques | 151 |
| Disposal on Ice | 151 |
| Substitutions | 151 |
| Other Alternatives | 152 |
| The "No Discharge" Alternative--A Case Study | 153 |
| References | 156 |
| Appendix A | 157 |
| Appendix B | 171 |

# Tables and Figures

Table


| | | |
|---|---|---|
| 1 | Offshore Wells in the United States | 12 |
| 2 | Mineral Composition of a Shale-Shaker Discharge from a Mid-Atlantic Well | 17 |
| 3 | Representative Fluid Compositions | 18 |
| 4 | Representative Metal Compositions | 20 |
| 5 | Drilling Fluid Components and Additives Used in the United States | 20 |
| 6 | How Specialty Additives Are Used to Solve Drilling Problems | 22-23 |
| 7 | Generic Fluid Systems (EPA Region II) | 28-29 |
| 8 | Characterizations of Field Drilling Fluids Used in the Joint Industry Mid-Atlantic Bioassay Program | 30-31 |
| 9 | Summary of Bioassay Results of Mid-Atlantic Generic Drilling Fluids | 32-33 |
| 10 | Average Discharge of Particulate Solids, Barium and Chromium from OCS Wells | 39 |
| 11 | Drainage Area and Water and Suspended Sediment Discharges of North America's Major Rivers | 40 |
| 12 | Estimates of Mass Emissions of Particulate Solids, Barium, and Chromium from OCS Drilling Discharges and from Rivers | 42 |
| 13 | Ocean Disposal of Various Wastes by Geographic Areas, 1973 to 1980 | 43 |
| 14 | Dilution and Dispersion of Discharge Plumes | 58 |
| 15 | Typical Sea Waves | 62 |
| 16 | Mass Balance of Total Excess Sediment Barium Surrounding Offshore Drilling Sites | 64-65 |
| 17 | Effect of Environmental Factors on Study Results | 67 |
| 18 | Acute Toxicity of Drilling Fluid Components to Estuarine and Marine Organisms | 77-78 |
| 19 | Summary of Results of Acute Lethal Bioassays With Drilling Fluids and Marine/Estuarine Organisms | 89-90 |
| 20 | Summary of Investigations of Sublethal and Chronic Effects of Drilling Fluids on Marine Animals | 91-95 |

| Table | | |
|---|---|---|
| 21 | Summary of Major Field Investigations of the Environmental Fate and Effects of Drilling Fluids and Cuttings Discharged to the Environment | 99 |
| 22 | Trace Metal Concentrations in Drilling Fluids from Different Sources | 106 |
| 23 | Summary of Biological Effects of Drilling Fluids and Drilling Fluid Ingredients on Marine Animals | 111 |
| 24 | Discharge Alternatives | 148 |
| 25 | Percentage of Time that Drilling Discharges Cannot be Transferred to Barges or Supply Boats | 150 |
| 26 | Estimated Costs of the "No Discharge" Alternative | 155 |
| A-1 | Chemicals Commonly Used in Drilling Fluid Lubricants | 164 |
| A-2 | Chemicals Commonly Used in Drilling Fluid Surfactants | 166 |
| A-3 | Chemicals Commonly Used in Drilling Fluid Emulsifiers | 166 |
| B-1 | Comparable Drilling Fluid Products by Tradenames | 172-180 |

| Figure | | |
|---|---|---|
| 1 | Drilling Fluid Circulation System | 15 |
| 2 | Idealized Jet Discharge | 52 |
| 3 | Variance as a Function of Time in Dye Diffusion Experiments | 55 |
| 4 | Barium Concentration as a Function of Transport Time | 59 |
| 5 | Relationship Between Current Speed, Particle Diameter, and Sediment Erosion, Transport, or Deposition | 68 |

# Summary, Conclusions, and Recommendations

The discharges made in drilling outer continental shelf (OCS) oil and gas wells have recently been the subject of research and public debate with regard to their potential effects on the marine environment. A lack of scientific consensus about the physical fates and biological effects of these discharges has led to actions contesting the permitting of some drilling discharges. At the request of the U.S. Department of the Interior, the National Research Council convened the Panel on Assessment of Fates and Effects of Drilling Fluids and Cuttings in the Marine Environment with the charge of establishing a credible technical basis for making resource management decisions.

This section presents the panel's summary, conclusions and recommendations. In conducting its assessment, the panel made numerous specific findings concerning fates and effects and inadequacies or gaps in available information. These are noted throughout the report. Many of these findings represent an incomplete understanding of basic oceanic or biological processes. The panel has taken these findings and limitations into account. Those which it considers to be the most salient and relevant are discussed in this section.

## THE USE AND COMPOSITION OF DRILLING FLUIDS

Drilling fluids are required in rotary drilling for oil and gas exploration and development to remove cuttings from beneath the bit, to control pressure in the well, to cool and lubricate the drill string, and to seal the well. There are no alternatives to using drilling fluids in this rotary drilling. Although drilling fluids are recirculated during drilling and sometimes can be held and reused in drilling multiple production wells, eventually they must be disposed of because of their contamination with suspended material or their loss of important properties or because of weight and space limitations on drilling vessels. Cuttings from the formation drilled are removed from the drilling fluid and must also be disposed of. Although drilling discharges can be barged ashore or to other sites at sea for disposal, cost and operational considerations favor onsite disposal, by either overboard discharge or shunting through a pipe to some depth. Land disposal is now required for certain drilling fluids (for example, oil-based drilling fluids), and in certain state waters (for example, some state waters of California and Alabama).

Drilling fluids used on the OCS are composed of bulk constituents and special purpose additives. The principal bulk constituents are water, barite (barium sulfate), clay minerals, chrome lignosulfonate, lignite, and sodium hydroxide. All of these constituents are nontoxic to marine organisms at the dilutions reached shortly after discharge. There is limited information on the compositions and quantities of additives in used fluids discharged on the OCS. Several common drilling-fluid additives, including biocides and diesel fuel (No. 2 fuel oil), are much more toxic to marine organisms than the bulk constituents.

Approximately two million metric tons (dry weight) of drilling-fluid components are discharged annually on the U.S. OCS, more than 90 percent of this amount in the Gulf of Mexico. Corresponding figures in the future will depend on government leasing policies, successes in exploration, and economic factors, but in the near future most drilling discharges are likely to occur in the Gulf of Mexico and off southern California and Alaska. Compared to the mass emissions of river sediments and those of municipal wastes and dredged material, the quantity of drilling fluids discharged in the ocean is small. For example, total particulate loading in the Gulf of Mexico from drilling fluids represents about 1 percent of that from the Mississippi River. Annual discharges of dredged material, of sludge, and of industrial wastes in U.S coastal waters exceed those of drilling fluids.

## THE CHEMICAL TOXICITY OF DRILLING FLUIDS

The first step in evaluating a material's potential harm to marine organisms and ecosystems is usually the acute lethal bioassay.* In this kind of test, organisms are exposed to graded concentrations of the material. Mortalities are recorded, and on the basis of these data the concentration causing 50-percent mortality after a predetermined exposure time (usually 96 hours) is estimated statistically and recorded as the median lethal concentration (LC50).

More than 96 percent of the whole drilling fluids tested in short-term experiments (from 44 to 144 hours) have LC50 values greater than 1,000 ppm and are classified as "slightly toxic" or "practically non-toxic" by the IMCO et al. (1969) characterization of toxicity (see Chapter 4). More than 98 percent of the tests that have used the suspended particulate phase of drilling fluids found their LC50 values greater than 10,000 ppm (in the range of "practically non-toxic").

---

*A bioassay is a quantitative determination of the concentration of a substance by its effect on an organism under controlled conditions.

This distribution of toxicities, representing over 70 drilling fluids and more than 60 species of marine organisms, indicates that most water-based drilling fluids are relatively nontoxic.

Fewer than 4 percent of the tests of whole fluids and only 2 percent of those using the suspended particulate phase found the substances "moderately toxic," that is, having LC50 values between 100 and 10,000 ppm. Most of this toxicity is probably attributable to the use of diesel fuel (No. 2 fuel oil) in the drilling fluids, but the fluids tested for toxicity have not always been fully analyzed chemically.

Acute toxicity bioassays are only the first step in hazard assessment. The results of these tests indicate the relative toxicities of used drilling fluids and the relative sensitivities of different species. They do not, for example, indicate sublethal signs of stress. Nor have such tests reproduced the exposure levels and intervals that characterize the dispersing plumes of discharged drilling fluids in the field.

Drilling fluids have recently been used in tests of sublethal toxicity. Such tests have measured changes in the growth and development of organisms in embryonic and larval stages and changes in the behavior of adults. In most cases, these effects are observed at concentrations of 10 to 1,000 ppm, about one-to-two orders of magnitude below LC50 values determined in acute bioassays. Expressed as an application factor of chronic to acute ratios, most species fall above a factor of 0.22; the highest ratio observed was 0.033. Unfortunately, the experimental designs of the tests of sublethal toxicity have also relied on exposure regimes that do not simulate the rapid dispersion of discharged drilling fluids or their movement along the bottom as measured in the field. Thus, hazard assessments using these biological data must extrapolate from them; yet there are no well-established relationships between responses and exposure intervals. The results of benthic microcosm experiments are also difficult to interpret. In these tests, responses to the chemical properties of drilling-fluid solids have not routinely been isolated from responses to physically altered substrates.

Predicting the effects of marine organisms' accumulation (through bioaccumulation) of substances in drilling fluids have relied on measurements of total tissue and body burdens and have not considered the organisms' mechanisms for sequestering and detoxifying contaminants. Nor have they taken into account whether contaminants are present at intracellular sites of toxic action. Furthermore, the potential increase of accumulated contaminate body burdens with increasing trophic levels has not been addressed, although research on other discharges containing the same metals suggests that the metals commonly found in drilling discharges are not biomagnified. The potential for biomagnification may be greater for organic compounds or organic complexed metals.

In toxicity tests, organisms from any one OCS region appear to be no more sensitive to drilling effluents than comparable ones from any other region, indicating that test results usually may be applied from one region to another. In addition, some nearshore organisms have shown sensitivities to drilling effluents similar to those exhibited by morphologically similar species from offshore areas; also, some species that have been tested are found both near and offshore. These results suggest that some nearshore species are appropriate surrogates for testing the effects of drilling effluents. It is desirable to tailor drilling-fluid regulations to take account or advantage of environmental conditions or to protect sensitive or valuable habitats, but there is no evidence that justifies different regulatory policies concerning the use of drilling-fluid additives in different geographic regions.

## THE PHYSICAL FATES OF DRILLING FLUIDS

Discharges of drilling fluids and cuttings into OCS waters take place in a wide range of marine environments, which vary greatly in water depth, ice cover, tidal and nontidal currents, waves, geological history, land runoff, and biotic characteristics. Thus, the physical fates of discharged drilling fluids and cuttings vary greatly.

On the continental shelf, approximately 90 percent of the particles in discharged drilling fluids, and almost all of the cuttings, settle rapidly, passing through a stage of convective descent until encountering the seabed or becoming neutrally buoyant. In addition to the main, or lower plume, a visible or upper plume is also formed. Most observations of water column fate have focused on the upper plume and on the dispersion of dissolved components. Based on observations of upper plumes, the plumes spread out at some depth appropriate to density characteristics and are rapidly dispersed by the turbulent diffusion characteristic of the ocean. Horizontal turbulent diffusion results in dilution of the plumes by a factor of 10,000 or more within an hour of release and even greater dilution of suspended components because of settling.

Although dilution may be inhomogeneous at thermoclines or pycnoclines, the high dilutions predicted in mathematical models take place in the field. Theoretical considerations and empirical observations yield the same values for dispersion rates in the water column. Given such rapid dilution within tens of meters of the discharge, toxic responses in organisms in the water column would be anticipated only if short-term exposures (of around one hour) result in acute effects at concentrations lower than 100 ppm. Although very few short exposure experiments have been conducted, longer term experiments (over 96 hours) have seldom identified lethal or sublethal effects at concentrations less than 100 ppm. Direct assessments of the effects on plankton and nekton in the water column have not been attempted and, given natural variability and the difficulty of sampling, are probably not feasible. Thus, even sublethal effects on pelagic biota moving past the point of discharge are confined to a very small area (within tens of meters) around the point of discharge. This finding suggests that restrictions on the dilutions or rates of discharges are not justified in most OCS areas.

At most depths typical of the continental shelf the majority of discharged fluids and cuttings are initially deposited on the seabed within 1,000 meters of the point of discharge. This material may persist as initially deposited or may undergo rapid or prolonged dispersion, depending on the energy of the bottom boundary layer. In high-energy environments, such as the tidally active Lower Cook Inlet in Alaska, the resuspensive and tractive dispersion of sedimented materials will take place very quickly. In relatively quiescent environments this dispersion will be slow and the fluids and cuttings may be physically or chemically detectable for a number of years. Storm events on the continental shelf probably control the accumulation of fluids and cuttings as much as any other environmental factor. In any case, the ultimate fates of the deposited materials depend on processes acting after deposition, which have not been treated in the conventional plume dispersion models.

The effects of drilling fluids and cuttings on benthic habitat, communities and organisms may be physical (burial or substrate change) and chemical (toxicity). In practice, it is difficult to separate physical and chemical effects based on either field surveys or laboratory experiments. Most laboratory experiments on the effects of drilling fluids on benthic organisms have not been very successful in mimicking realistic exposure conditions. Effects on benthos have been observed in the field, under low to moderate energy regimes, within 1,000 meters of the discharge point. Only one study has yet described environmental changes over time after drilling operations ceased; while the fauna had been altered, recovery was nearly complete within one year. Because the effects of drilling discharges are probably largely physical, recovery times should be similar to those following other physical seabed disturbances. These times vary widely; recovery may take weeks in frequently disturbed shallow-water communities, several months to several years in continental shelf communities, and many years on the continental slope and in deep sea. The resuspensive transport of deposited drilling-fluid components may produce effects beyond the area of immediate burial, but at the same time it reduces the concentrations of potentially toxic substances. As the material disperses, organisms that feed at the sediment-water interface may nonetheless be exposed to higher concentrations of such substances than bulk analysis of sediments would suggest.

Shunting drilling discharges to the near-bottom, as an alternative to surface disposal, may increase the exposure of benthic organisms to wastes. It may be effective, however, in restricting wastes from topographic rises with sensitive biota like reef corals. In contrast, surface discharges ensure dispersion and limit the duration and amount of organism exposure. Predilution of such discharges is generally unnecessary given the speed with which they are diluted, except possibly in low-energy or shallow-water environments.

The long-term benthic effects of drilling discharges from multiple wells during intensive exploration or development are difficult to distinguish from the effects of other discharges and activities (including oil and gas production) on the continental shelf and from natural variations. Comprehensive studies of these various effects are not available. Results of platform monitoring studies have demonstrated spatially limited effects on the benthos. However, these effects cannot be directly ascribed to discharges of drilling fluids. Long-lived communities, which are characteristic of hard substrate epibiota, may be particularly susceptible to long-term effects if they are exposed to large concentrations of deposited fluids and cuttings, but many of these communities are not very likely to accumulate such materials unless the materials are deposited directly on them.

## CURRENT KNOWLEDGE OF DRILLING-FLUIDS FATES AND EFFECTS

The information base for assessing the fates and effects of drilling discharges in OCS waters has some notable deficiencies, many of which pertain equally to the effects of other pollutants in the coastal ocean. These deficiencies include variable quality of research, limits to the realism and relevance of laboratory experiments, difficulties in unequivocally ascribing effects observed in field studies to given causes, and a poor understanding of ecosystem processes. These limitations do not invalidate most of the results that have been produced, but must be taken into account in interpreting them. Our knowledge of the fates and effects of drilling fluids and cuttings is not notably inferior to that of the fates annd effects of dredged materials and other wastes dumped in the ocean, even though the latter have been studied considerably longer.

Our understanding of the fates and effects of drilling discharges in the marine environment is limited more by the state of our general understanding of marine pollution than by specific deficiencies in our knowledge of drilling fluids and cuttings. Our understanding of this narrow problem may be advanced most rapidly by conducting research on the broader topics of the accumulation and transfer of materials in the marine environment. With this understanding of where research emphasis should be placed, the panel concludes that extensive further research focused specifically on the fates and effects of drilling fluid discharges is not needed.

Any additional research on drilling fluids should include acute, sublethal, and chronic bioassays using techniques and contaminant exposures that reflect actual discharge and exposure conditions, field studies that take into account inventories and chemical analyses of discharges, and studies of resuspensive transport of particulate contaminants.

The panel's review of existing information on the fates and effects of drilling fluids and cuttings on the OCS shows that the effects of individual discharges are quite limited in extent and are confined mainly to the benthic environment. These results suggest that the environmental risks of exploratory drilling discharges to most OCS communities are small. Discharges from oil and gas field development drilling introduce greater quantities of material into the marine environment over longer periods of time. Results of field studies suggest that the accumulation of materials from these longer-term inputs is less than additive and therefore the effects of exploratory drilling provide a reasonable model for projecting the effects of development drilling. Uncertainties regarding effects still exist for low energy depositional environments, which experience large inputs of drilling discharges over long periods of time.

To minimize effects, care needs to be exercised in the following:

- Discharges should be prevented from burying particularly sensitive benthic environments, especially hard substrate epibiota, which are not exposed to significant natural sediment flux.
- The use of more toxic additives, such as diesel fuel (No. 2 diesel oil), should be monitored or limited. Fluids that show significant toxicity should be analyzed chemically to determine their toxic components.

# 1 Introduction

This report seeks to answer two questions:

- Are drilling fluids and cuttings as they are released into the outer continental shelf (OCS) toxic to marine organisms or do they cause deleterious sublethal responses in these organisms that may adversely affect the ecosystem, or are they innocuous?
- Are the heavy metals or organic materials in drilling fluids or cuttings bioaccumulated or biomagnified so that they are harmful to organisms or to those who consume them, including man?

Chapter 2 of the report, "Drilling Discharges," provides an overview of offshore drilling, and of the use, composition, and chemistry of drilling fluids and cuttings. It describes the regulation of drilling discharges and examines the quantities and frequencies of these discharges and their components in relation to other inputs to the marine environment.

Succeeding chapters review what is known about the fates and effects of drilling fluids and cuttings in the marine environment. Chapter 3, "The Fates of Drilling Discharges," discusses the transport forces of the ocean, the behavior of dissolved and particulate materials in seawater, and the physical fates of drilling fluids and cuttings in the marine environment. Chapter 4, "Biological Effects of Drilling Discharges," summarizes and critically evaluates the scientific literature on the toxicities of drilling fluids and on the impacts on OCS ecosystems of discharging used drilling fluids and cuttings.

After reviewing the available information on these topics the report discusses the limitations in using this information for decision making. Chapter 5 reviews the adequacy and applicability of the available information. It reviews considerations in conducting and interpreting laboratory evaluations of toxicity and field studies, problems of extrapolating from laboratory to field and between geographic regions, and the topics of bioaccumulation and the long-term fates and effects of drilling discharges.

The final chapter considers the operations, cost, and risk of disposing drill discharges overboard.

The conclusions and recommendations of the panel appear in the summary.

# 2
# Drilling Discharges

## OFFSHORE OIL AND GAS DRILLING AND DEVELOPMENT

Since 1947 nearly 22,000 wells have been drilled on the OCS in exploring for and developing oil and gas resources. Table 1 indicates that those in the Gulf of Mexico (offshore[1] Florida, Louisiana, and Texas) account for 83.2 percent of all offshore wells; those offshore Louisiana alone account for 73.3 percent. Two-thirds of all offshore wells (67.1 percent) have been drilled in federal waters, although this percentage varies widely by geographic region, from 100 percent in the offshore Atlantic to 5 percent offshore Alaska. Exploratory wells account for 24.6 percent of the wells in federal waters and for 23.2 percent of those in state waters. These figures also vary widely by region: all wells drilled in the Atlantic, where there have been no commercial discoveries, have been exploratory; 91.2 percent of wells drilled offshore California, where offshore development began in the 1890s, have been development wells.

OCS oil and gas production now accounts for 8 percent of domestic oil production and 24 percent of domestic gas production (Minerals Management Service, 1982). The U.S. Geological Survey estimates that as much as 41.3 percent of the nation's undiscovered recoverable oil and 28.1 percent of its natural gas lie offshore (Dolton et al., 1981).

In the future, major new discoveries of oil and gas are more likely to occur offshore than on land, because the OCS has been less completely explored. Such large discoveries, like those recently made offshore California and Alabama, are also more cost-effective to develop than multiple smaller discoveries (which characterize the majority of past Gulf of Mexico developments). Thus, the sites of future oil and gas development are likely to be those areas that have not yet been thoroughly explored, and that have geologic potential for large accumulations of oil and gas (Edgar, 1983). Offshore Alaska is one area that meets these criteria. By 1990, 22 percent of domestic oil is expected to come from future discoveries, 45 percent by the year 2000 (Palmer and Kelly, 1983).

---

[1] The term "offshore" is used in this report to refer to state and federal offshore lands together, "OCS" to federal lands only.

TABLE 1 Offshore Wells in the United States[a,b]

| | Exploratory | Development | Total |
|---|---|---|---|
| Alaska | | | |
| State | 80 | 281 | 361 |
| Federal | 19 | - | 19 |
| Total | 99 | 281 | 380 |
| California | | | |
| State | 161 | 3,185 | 3,364 |
| Federal | 145 | 299 | 444 |
| Total | 306 | 3,484 | 3,790 |
| Oregon | | | |
| Federal | 8 | - | 8 |
| Washington | | | |
| State | 2 | - | 2 |
| Federal | 4 | - | 4 |
| Total | 6 | | 6 |
| Florida | | | |
| State | 15 | - | 15 |
| Federal | 9 | - | 9 |
| Total | 24 | | 24 |
| Louisiana | | | |
| State | 977 | 2,904 | 3,881 |
| Federal | 3,186 | 12,134 | 15,320 |
| Total | 4,163 | 15,038 | 19,201 |
| Texas | | | |
| State | 762 | 252 | 1,014 |
| Federal | 888 | 656 | 1,544 |
| Total | 1,650 | 908 | 2,558 |
| Atlantic | | | |
| Federal | 21 | - | 21 |
| Total | | | |
| State | 1,997 | 6,622 | 8,619 |
| Federal | 4,280 | 13,089 | 17,369 |
| State and federal | 6,277 | 19,711 | 25,988 |

[a]Cumulative through 1981. New wells are now drilled on the OCS at the rate of approximately 1,000 per year.
[b]Offshore wells are defined as those beyond natural shorelines.

SOURCE: Adapted from American Petroleum Institute (1982).

## CHARACTERISTICS AND FUNCTIONS OF DRILLING FLUIDS

Commercial oil and gas exploration and production wells on the OCS are drilled with rotary equipment. In rotary drilling, the well is drilled by a rotating bit to which downward force is applied. The bit is fastened to and rotated by a hollow drill stem made of pipe, through which drilling fluid is circulated.

Drilling fluids are essential to drilling operations, performing the following major functions:[2]

- Removing cuttings from beneath the bit and transporting them to the surface where they can be separated from the drilling fluid for disposal
- Preventing formation fluids from flowing into the wellbore by maintaining a hydrostatic pressure in excess of the fluid pressure in the formation
- Coating the borehole wall with an impermeable filter cake to prevent fluid loss in permeable formations
- Having sufficiently high gel properties to suspend cuttings and fluid solids when circulation is interrupted
- Helping to support the weight of the drill string
- Lubricating and cooling the drill bit and drill string
- Having properties that do not interfere with the accurate geological evaluation of the formation or the production of oil and gas.

Drilling fluids are classified as either water-based or oil-based, depending on their principal liquid-phase component. Consistent with the scope of this report, this section is concerned with water-based drilling fluids.

---

[2] In completing wells for production and in workover operations (periodic maintenance) special fluids may be used:

- A packer fluid may be placed in the well to counter formation pressures over a long period of time.
- In workover operations, special drilling fluids may be circulated continuously for the limited duration of the operation.
- Special fluids may be used in stimulation procedures such as fracturing. (In this operation, fluids can sometimes be displaced into the formation. The potential for their discharge exists when the well is returned to production.)

Some fluids used in these operations, such as seawater, are innocuous and discharged routinely. Others are oil-based and disposed of onshore or are fluids containing high concentrations of soluble salts (e.g., $CaCl_2/CaBr_2$ and $CaCl_2/CaBr_2/ZnBr_2$). These discharges are quite small relative to other drilling discharges and are beyond the scope of this report.

A schematic diagram of a drilling fluid circulation system is shown in Figure 1. The fluid's components are added through the hopper and mixed in the tanks. The fluid is then pumped from the tanks down the drill string and through the bit. It sweeps the crushed rock cuttings from beneath the bit and carries them back up the annular space between the drill string and the borehole or casing to the surface. This permits drilling to continue and is the fluid's most important function.

## DISCHARGES OF DRILLING FLUIDS[3]

After the drilling fluid has circulated through the well and has returned to the surface, it is passed through solids control equipment to remove the formation drill solids (cuttings). The solids control equipment is an integrated system that consists of shale-shaker screens[4] that remove the coarse particles and hydrocyclones[5] that remove the sand and silt fractions from the fluid. The drill solids separated by the solids control equipment are discharged to the ocean. This type of discharge is continuous in that it occurs while drilling is in progress. Typically, these discharges occur about half the time the rig is on location.

The rates of this type of discharge vary from about 1 to 10 bbl/h[6] (Ayers, 1981). The higher number is more characteristic of the shallow part of the hole when drilling is fast and the bit diameter is large. Over the life of a well, some 3,000 to 6,000 bbl of wet solids are discharged from the solids control equipment (Ayers, 1981).

After the fluid passes through the solids control equipment and the solids are separated, it is returned to the tanks for recirculation. At this point another type of discharge may be required. The solids control equipment cannot remove the fine clay

---

[3]This section presents information on discharges from wells. A discussion of mass loading--cumulative discharge quantities--appears in the the final section of this chapter.

[4]A series of trays with sieves that vibrate to remove cuttings from the circulating fluid. The size of the openings in the sieves is selected to match the size of the solids in the drilling fluid and the anticipated size of cuttings (Petroleum Extension Service, 1979).

[5]A centrifugal device used to remove fine particles of sand from drilling fluid. It operates on the principle of a fast-moving stream of fluid being put into a whirling motion inside a cone-shaped vessel (Petroleum Extension Service, 1979).

[6]A measure of volume for petroleum products. One barrel (bbl) equals 42 U.S. gallons. One $m^3$ equals 6.2897 bbl (Petroleum Extension Service, 1979).

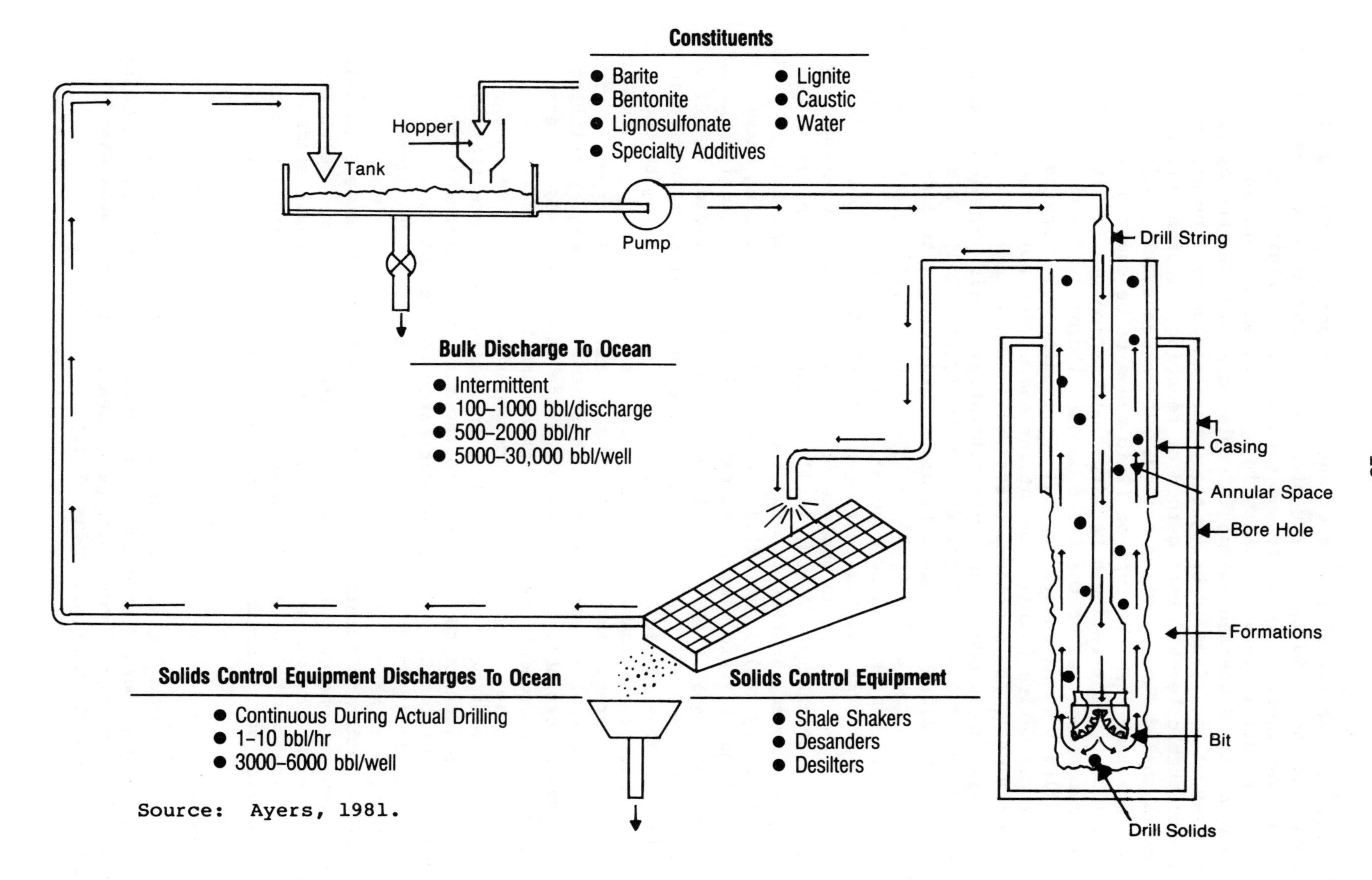


Source: Ayers, 1981.

FIGURE 1 Drilling Fluid Circulation System

and colloidal particles[7] that are generated in drilling through formations. As the fluid is recirculated the concentration of these fine particles continues to increase and eventually the fluid becomes too viscous for further use. At this time, a portion of the fluid is discharged and the discarded volume is replaced with water and appropriate quantities of additives to bring the concentration of fine solids back to an acceptable level. This method of reducing the fine solids in the system is called "the dilution method." Less frequently, bulk discharges are made when the type of fluid needs to be changed as when the bit will be penetrating a particular formation or when the rheological properties of the fluid become altered. (The chemistry of drilling fluids is discussed below.) It is also necessary to discharge the entire fluid system at the end of drilling each exploratory well, and sometimes after drilling development wells.

Bulk discharges occur only intermittently. Their volumes normally range from 100 to 1,000 bbl per discharge (Ayers, 1981). A small volume, 100 to 200 bbl, is usually discharged every 1 to 3 days (Ayers, 1981). A discharge of 1,000 bbl is typical on completing a well or when the fluid system must be changed for some reason.

The rate of bulk discharges ranges from 500 to 2,000 bbl/h (Ayers, 1981). Over the life of an exploratory well, some 5,000 to 30,000 bbl of fluid are discharged (Ayers, 1981). Because development wells are normally shallower, smaller in diameter, and require less time to drill than exploratory wells, less fluid is discharged in drilling them.

The volume of fluid discharged ranges widely. The dilution method is an efficient way to control the concentration of colloids and fine particles in low-density fluids, which contain a minimum of barite and additives. (These fluids are adequate for shallow drilling through competent rock formations.) Discharge volumes will usually be high for this type of system, since the bulk of the material discharged is water and the fluids cost is low. On the other hand, high-density drilling fluids have appreciable quantities of barite and additives, and are expensive. For economic reasons, it is desirable to minimize the bulk discharge of these fluids. This is accomplished by the more extensive use of solids control equipment and by increasing the concentration of chrome lignosulfonate to deflocculate the fine clay particles and to reduce fluid viscosity. Thus, the discharge volumes of such high-density fluids are low compared to those of less expensive low-density systems.

The variation in quantity of discharged material from well to well is much less if one considers only the quantity of solids--everything but water--that is discharged. About 1,000 $m^3$ (2,000 tons) of dry solids (formation solids and fluid additives) are discharged both in bulk and from solids control equipment over the life of a typical

---

[7]A colloid is a liquid mixture in which the particles of one substance are dispersed in another in a continuous phase without being dissolved. The size of colloidal particles is in the approximate range of 10-5,000 Angstroms (Adam, 1956).

exploratory well (Ayers, 1981). The quantity of discharges from development wells is likely to be as much as 25 percent less than that of discharges from exploratory wells (Ayers, 1983). Fluid components account for about half the quantity of discharges in dry weight and formation solids for the other half.

## THE COMPOSITIONS OF DISCHARGES

The discharges from solids control equipment and those made in bulk have different compositions. The first contain primarily formation solids, and the second fluid components. Table 2 gives the mineral composition of a shale-shaker discharge. This sample and others discussed below are representative of solids discharges from drilling operations in that the primary constituents are naturally occurring

TABLE 2 Mineral Composition of a Shale-Shaker Discharge From a Mid-Atlantic Well[a]

| Mineral | Percentage by Weight (Dry Basis) |
|---|---|
| Barium Sulfate | 3 |
| Montmorillonite | 21 |
| Illite | 11 |
| Kaolinite | 11 |
| Chlorite | 6 |
| Muscovite | 5 |
| Quartz | 23 |
| Feldspar | 8 |
| Calcite | 5 |
| Pyrite | 2 |
| Siderite | 4 |

[a]Sixty-five percent solids, density 1.7 $g/cm^3$.

SOURCE: Adapted from Ayers, Sauer, Meek, and Bowers (1980).

clay and quartz minerals. This sample was obtained from a well drilled in the mid-Atlantic region, about 100 miles East of Atlantic City, New Jersey. The small amount of barium sulfate results from barite particles that have adhered to the cuttings particles. The montmorillonite clay comes from both added bentonite and from formation clays. The remaining material represents the formation being drilled at that time and consists primarily of clays, quartz, and low concentrations of calcite, pyrite, and siderite. This particular shale-shaker sample contained 65 percent solids and 35 percent water. The amount of water in these discharges ranges from 20 to 50 percent.

The compositions of drilling fluids vary with both the depth and the location of the well. In the shallow portion of the hole the fluid used usually consists of low concentrations of bentonite and sodium hydroxide in seawater ("spud mud"). As hole depth increases, the system may be converted to fresh water with more bentonite, lignite, lignosulfonate, and barite added. Also, if problems in drilling occur, specialty chemicals may be required (see Table 6). The vast majority of fluids discharged on the OCS are like the two compositions shown in Table 3. One is a low-density and the other a high-density fluid. The high-density sample represents the final composition of the fluid used in a well in the Gulf of Mexico, and the low-density one represents a fluid used in the late stages of drilling a mid-Atlantic well. The high-density fluid weighs 2.1 g/cm$^3$ and contains 62 percent barite

TABLE 3 Representative Fluid Compositions

| | Concentration (wt%) | |
|---|---|---|
| Component | Low Density[a] (1.19 g/cm$^3$) | High Density[b] (2.09 g/cm$^3$) |
| Barite | 15.0 | 62.0 |
| Low gravity solids | 6.5 | 5.9 |
| Chrome lignosulfonate | 1.0 | 0.9 |
| Lignite | 1.0 | 0.9 |
| Inorganic salts | 0.7 | 0.5 |
| Water | 75.8 | 29.8 |

[a] pH 11.4.
[b] pH 12.4.

SOURCE: Adapted from Ayers (1981).

and 30 percent water. The low-density fluid weighs 1.2 g/$cm^3$ and contains 15 percent barite and 76 percent water. The concentrations of other ingredients in the two fluids--low-gravity solids, chrome lignosulfonate, and lignite--are similar. The low-gravity solids are bentonite clay and formation solids.

Trace metals in drilling discharges originate from both formation solids and fluid additives. Representative metal concentrations for a shale-shaker sample and a fluid sample are shown in Table 4. These samples were taken from a well in the mid-Atlantic. The presence of barite causes the barium concentration to be much higher than that of any other metal. Chromium also occurs in concentrations higher than those normally seen in formation solids or sediments. The chromium comes from the additive chrome lignosulfonate. Both barium and chromium concentrations are higher in the fluid than in the shale-shaker sample because these metals come from fluid additives, and only a small quantity of fluid additives adheres to the cuttings when they are screened out.[8] The other metals shown in Table 4 are present in concentrations comparable to those normally found in formation solids or sediments.

## COMPONENTS OF WATER-BASED DRILLING FLUIDS

Five major components (barite, clays, lignosulfonate, lignite, and caustic soda) account for over 90 percent of the solid components of water-based drilling fluids (Perricone, 1980), as is illustrated in Table 5. Appendix A provides a review of the functional components of drilling fluids. These five components and water account for over 98 percent of the mass (or volume) of drilling fluid discharged to the OCS. These components, in decreasing order of use, are the following:

- Barite, a mineral containing 80 to 90 percent barium sulfate, which is used to increase the density of the drilling fluid to control formation pressures. In some cases, concentrations as high as 700 lb/bbl may be used. Depending on its source, barite may contain low concentrations of quartz, chert, silicates, and other minerals and also trace levels of metals.
- Bentonite, the clay most commonly used in drilling fluids. Sodium montmorillonite clay in concentrations of 60 to 80 percent is the predominant ingredient. Silica, shale, calcite, mica, and feldspar are common impurities in bentonite deposits. Bentonite is used to maintain the rheological properties required to remove the cuttings from beneath the bit and carry them to the surface. Bentonite also

---

[8] The quantity of additives that adheres to cuttings depends on the depth of water in which the cuttings are discharged (residence time), and the mixing energy of the water column. As much as 20 percent of additives may adhere to cuttings in a shallow water, low energy environment.

TABLE 4 Representative Metal Compositions

| Metal | Concentration (mg/kg) | |
|---|---|---|
| | Shale Shaker[a] | Fluid[b] |
| Barium | 3,160 | 37,400 |
| Chromium | 44 | 191 |
| Cadmium | $<2$ | $<1$ |
| Lead | 10 | 3 |
| Mercury | $<1$ | $<1$ |
| Nickel | 15 | 4 |
| Vanadium | 11 | 5 |
| Zinc | 80 | 50 |

[a] 77.1 percent solids, 1.9 $g/cm^3$.
[b] 21.0 percent solids, 1.16 $g/cm^3$.

SOURCE: Adapted from Ayers, Sauer, Meek, and Bowers (1980).

TABLE 5 Drilling Fluid Components and Additives Used in the United States

| Component | Percentage of Total |
|---|---|
| Barite | 63.0 |
| Clays | 24.0 |
| Lignosulfonate | 2.0 |
| Lignite | 1.5 |
| Sodium hydroxide | 1.5 |
| Other additives[a] | 8.0 |

[a]Special additives to oil-based drilling fluids are included in this estimate.

SOURCE: Adapted from American Petroleum Institute (1978).

prevents fluid loss by providing filtration control while drilling through permeable zones. The concentration of bentonite in drilling fluids normally ranges from 5 to 35 lb/bbl.

- Lignosulfonates, which are normally used in drilling fluids in concentrations ranging from 1 to 15 lb/bbl. Lignosulfonates are derived from the sulfite pulping of wood chips to produce paper and cellulose. Chrome lignosulfonates, the most widely used deflocculant in drilling fluids, are prepared by treating lignosulfonate with sulfuric acid and sodium dichromate. Sodium dichromate oxidizes the lignosulfonate and cross-linking occurs. Hexavalent chromium introduced by the chromate is reduced during the reaction to the trivalent state and complexes with the lignosulfonate. Lignosulfonates control viscosity in water-based drilling fluids by acting as thinning agents or deflocculants for clay particles. The chrome appears to bind onto the edges of clay particles at high downhole temperatures, reducing the formation of colloids (Skelly and Dieball, 1970).
- Lignite (soft coal) which is used in drilling fluids as a clay deflocculant and to control filtration rate. The concentration of lignite in drilling fluids normally ranges from 1 to 15 lb/bbl. Most of the drilling-grade lignite, leonardite, is mined in North Dakota. The chief constituent of this naturally occuring oxidized lignite is humic acid.
- Sodium hydroxide (caustic soda), which is normally used in drilling fluids in concentrations sufficient to maintain a pH of 9 to 12. A pH greater than 9.5 is needed to obtain maximum deflocculation from the chrome lignosulfonate and to keep lignite in solution. A basic pH also lowers corrosion rates and provides protection against possible hydrogen sulfide contamination by suppressing microbial growth.

A large number of other additives are available for use in water-based drilling fluids (American Petroleum Institute, 1978). These additives, which have been formulated to meet specific needs, range in complexity from simple inorganic salts to organic polymers of high molecular weight. Typically, only a few are used on any one well, and they are used in low concentrations (Moseley, 1981). Table 6 gives the operating objectives of the most frequently used and environmentally significant additives in water-based drilling fluids, and indicates their ranges of concentration and frequencies of use.

Water-based drilling fluids and drill cuttings sometimes contain quantities of hydrocarbons (usually diesel fuel [No. 2 fuel oil]) in greater than trace amounts. This occurs when diesel fuel is added to the fluid system to reduce torque and drag. As much as 2 to 4 percent diesel may be added to the bulk fluid system to improve lubricity (a relatively common operating practice in the Gulf of Mexico). A standard technique for freeing the drill pipe should it become stuck, is to pump a "pill" of diesel fuel or oil-based drilling fluid down the drill string and "spot" it in the annulus area where the pipe is stuck. The pill may or may not be kept separate from the bulk drilling fluid system, recovered, and disposed of onshore. Even when the pill

TABLE 6 Special Additives and Their Uses

| Additive | Operating Objective | Concentration (lb/bbl) | Frequency of Use[a] |
|---|---|---|---|
| Sodium bicarbonate | Eliminate excess calcium ions due to cement contamination by precipitating calcium as calcium carbonate. | 0.1-4 | Very common |
| Sodium chloride | Minimize borehole washout in salt zone by preventing dissolution of salt formation. | 10-125 | Rare |
| Ground nut shells, mica, or cellophane | Minimize loss of drilling fluid to the formation by adding material to plug the "thief" zone. | 5-50 | Common |
| Cellulose polymers or starch | Counter thick, sticky filter cake; decrease filtrate loss to formation. | 0.25-5 | Very common |
| Aluminum stearate or alcohols | Minimize foaming. | 0.05-0.1 | Common |
| Sodium chromate | Reduce viscosity increase in high temperature wells; aid deflocculation of lignosulfonate. | 0.1-2 | Rare |
| Diesel, vegetable, or mineral oil lubricant | Reduce torque and drag on the drill string by preventing it from sticking. | 2-50 | Common |

| | | | |
|---|---|---|---|
| Pill of oil-based spotting fluid | Counter differential pressure sticking of drill string. (Pill is placed downhole opposite contact zone to free pipe. After pipe is free, the oil-contaminated mud is collected and may or may not be discharged to the ocean depending on operational circumstances.) | 100-300[b] | Common |
| Paraformaldehyde bactericide | Retard bacterial degradation in polymer starch fluid systems; | 0.2-2 | Very common |
| | prevent casing string corrosion in development drilling when added to fluid left behind the casing. | 0.2-2 | Very common |
| Zinc compounds | Counter hydrogen sulfide contamination by precipitating sulfides. | 0.5-5 | Common |
| Potassium Chloride | Prevent shale swelling and sloughing; improve wellbore stability. | 20-95 | Rare |
| Biopolymer | Provide viscosity in drilling fluids with high salt concentrations. | 0.2-2 | Rare |
| Asbestos[c] | Improve solids-carrying capacity; lift formation drill solids out of the hole. | 1-10 | Very rare |

[a] Characterizations are expert judgments, based in part on the quantities of additives sold in 1978 and the concentrations of additives used.
[b] Concentration of oil in the pill of fluid.
[c] The use of asbestos is prohibited in most OCS regions under EPA's NPDES program.

SOURCE: Adapted from American Petroleum Institute (1978) and Moseley (1981).

is recovered, a small amount of diesel fuel from the pill may become mixed with the bulk drilling fluid. Discharges of water based drilling fluids containing diesel fuel are not prohibited in the Gulf of Mexico provided the discharge does not cause an oil sheen on the water surface or an oily sludge on the seafloor.

## THE CHEMISTRY OF DRILLING FLUIDS

The chemistry of drilling fluids is complex because of the diversity of components that may be used and the high temperatures and pressures that may be encountered at depth. As slurries, drilling fluids have some attributes of liquids, yet tests used in aquatic chemistry are often inappropriate for them because of their high solids contents. To confound the chemists further, soil chemistry procedures are often not appropriate because of the high water content of drilling fluids. Further complicating matters, the high temperatures and pressures encountered in some wells can dramatically affect chemical equilibria.

Water-based drilling fluids are colloids, suspensions of fine particles in solution. Understanding their chemistry begins with understanding the behavior of colloidal clays in water. Organic colloids are also present in drilling fluids, and, like inorganic colloids, are chemically active. The particle sizes of these chemical groups are so small that properties like viscosity and sedimentation velocity are controlled by surface chemistry phenomena. Furthermore, the surficial layers of the clay particles, and in some cases organic molecules, are charged. Clays particularly have high surface area to volume ratios and therefore high charge to mass ratios. End-to-end, side-to-side aggregations that form as a result are the basic mechanisms of flocculation and viscosity, and are essential to understanding thinning mechanisms. Gray et al. (1980) provide several good sections on clay chemistry and discuss colloidal interactions, as does van Olphen (1977).

With the exception of electrochemical changes in clays, most drilling-fluid solids do not undergo chemical changes as a result of the temperatures and physical conditions that occur in drilling. Even so, maintaining rheological properties with increasing depth of drilling is a major technical challenge because of the increased tendency of clays to flocculate at the higher temperatures encountered.

Carney and Harris (1975) grouped drilling-fluid additives according to their thermal stability. Their discussion of thermal degradation of lignosulfonates and the work of Skelly and Kjellstrad (1966) are important to understanding drilling-fluid chemistry. Clay particles, which in slurry form provide lubrication, tend to aggregate. To retard this process, the drilling fluid may be diluted with water if it contains a minimum of solids; in heavier fluids, chemical thinners like ferrochrome lignosulfonate may be added (McAtee and Smith, 1969). Chrome lignosulfonate adsorbs on the edges of clay particles and prevents them from flocculating (Skelly and Dieball, 1970). At high temperatures, higher concentrations of chrome lignosulfonate are required (compare drilling fluids 7 and 8 in Table 7) because the chrome lignosulfonate undergoes thermal degradation (some polymerization occurs in this reaction, releasing carbonates, bicarbonates, and sulfates). When the concentration of fine particles becomes so great that flocculation can not be controlled through the use of additives, the drilling fluid must be replaced.

## DRILLING-FLUID COMPONENTS AS COMMERCIAL PRODUCTS

The drilling-fluids industry has grown to be a major oilfield service industry worldwide. Four U.S. companies control approximately 90 percent of the world market. In the United States, smaller companies are better able to compete and may capture 25 percent of domestic sales (Escott and Walker, 1981). Each of the four major companies is integrated to the extent that it mines, processes, packages, distributes, stores, and delivers to the well site the major bulk products (e.g., barite and bentonite). These companies also provide onsite consulting, including operating recommendations and product information and testing.

Some of the components of drilling fluids (e.g., caustic soda) are commodity chemicals widely produced and used. Others are specialty products developed for and used exclusively in drilling fluids. While the commodity chemicals do not represent a large number of available drilling-fluid products, they do represent the major part of drilling fluid additives by weight and are present in nearly all drilling fluids.

The significance of the distinction between commodity chemicals and specialty products relates to the information available on chemical composition. Chemical information on commodity chemicals is widely available and usually appears in detail on product containers or technical data sheets issued by the responsible company. Chemical information on specialty products may or may not be as specific, depending on the product's patent status. Information on patented products and systems is usually available and in the public domain. Products not patentable or for which patents have not been issued are usually described in less chemical detail. However, chemical family names at least are available, and more specific data may be released if required for product registration or approval.

If regulated hazardous substances are included in the product, these compounds will be listed in the required terms on the container, in the product literature, or on the Material Safety Data Sheet (Occupational Safety and Health Administration (OSHA) Form 20). Complete information on the composition of bactericides is required under the Federal Insecticide, Fungicide, and Rodenticide Act. With these exceptions, however, neither chemical formulas nor manufacturing processes are described. This allows the manufacturing company to maintain a stronger market position with regard to a product or system.

The development of drilling fluid products is driven by the same forces that drive the development of other specialty chemicals--availability of resources, proven product performance, proven marketability, available technology, and favorable return on investment

(McGuire, 1973). Their constraints are similar--competition, changing markets, government regulations, and customer demands. Drilling-fluid service companies undertake the majority of product-related research and development, but much is also performed by major oil companies, chemical companies, and academic and research institutions. As in the chemical industry generally (Ashford and Heaton, 1979), government regulations protecting the health of the worker, the public, and the environment have caused the development of additional health and safety data, product substitutions or modifications, and removals of products from the market. These regulations have prompted the development of some products that are designed to be not only functional but more "environmentally acceptable" (Jones et al., 1980), for example, by substituting mineral or vegetable oil for diesel.

In part because of the numbers of commercial chemicals used in drilling fluids, drilling fluid companies seek to protect their market positions through the use of trade names. A list of drilling-fluid components (Wright and Dudley, 1982) suggests there are thousands of such components, but the profusion of trade names makes the list considerably redundant (American Petroleum Institute, 1978). Appendix B lists functionally equivalent products of the four leading drilling-fluid service companies.

## TRENDS IN OPERATING PRACTICES: GENERIC DRILLING FLUIDS

While numerous products are available for use in drilling-fluid systems (Wright and Dudley, 1982), in practice the number of generic chemicals (as opposed to trade-name products) is limited. In 1978 the Offshore Operators Committee (OOC) and the U.S. Environmental Protection Agency (EPA), Region II, capitalized on this uniformity by developing the generic drilling-fluid concept. Eight basic drilling-fluid systems were designated that encompass most drilling fluid types commonly used offshore. These systems are described in Tables 7 and 8 (Ayers, Sauer, and Anderson, 1983). The impetus behind identifying and using these categories is to address the toxicity of drilling discharges under Sec. 403 of the Clean Water Act by providing EPA with an understanding of, and control over, drilling-fluid formulations and discharges without requiring operators to perform redundant bioassays and chemical tests for every permitted discharge. The concept also has been adopted in EPA Regions I (for Georges Bank), II (Baltimore Canyon region), and IX (for California), and is being considered for use in EPA regions III (mid-Atlantic), IV (eastern Gulf of Mexico), VI (Western Gulf of Mexico), and X (Alaska).

The eight generic drilling fluids in the tables were identified by reviewing permit requests in EPA Region II and selecting the minimum number of fluid systems which would cover all of the prospective permits. The eight generic fluids contain primarily major components and do not consider specialty additives. Therefore, lists of frequently used additives have also been developed in each region. EPA has required that bioassays of both generic drilling fluids and additives be completed as a condition of their initial approval (see, for example, Table 9). Drilling operators may use an additive that is not on the approved list if data are submitted to EPA prior to its use on its chemical composition, rates of use, and toxicity. Such special discharges are approved case by case. Once bioassay tests on an additive have been completed, the additive may be added to the approved list, provided it does not significantly alter drilling-fluid toxicity (Jones and Hulse, 1982). All permits provide for "emergency use" of specialty additives.

Two or three generic drilling fluids may be used in a well. For example, initial drilling is usually conducted with a spud fluid. As drilling progresses, increasing amounts of weighting agents and thinners are added. Thus, one drilling program may call successively for a spud fluid, a lightly treated lignosulfonate fluid, and then a lignosulfonate freshwater fluid. Another program may call for a potassium chloride (KCl) system. This system makes extensive use of polymers to control viscosity, with bactericides sometimes added to the system to keep the polymers from degrading (IMCO Services, 1978).

Some generic fluids are saltwater fluids, others are freshwater. Saltwater fluids, commonly with concentrations of salt greater than 10,000 ppm, are used when drilling salt sections that would collapse if freshwater fluids were used, when resistivity control is needed, when drilling through bentonite shales, or when fresh water is not available in large quantities. The addition of saltwater to freshwater fluids increases viscosity and reduces gel strength with resulting loss of fluid. Certain properties are more difficult to maintain in saltwater than in freshwater fluids. Saltwater fluids require more dispersants and deflocculants to control viscosity and to maintain gel strength. For these reasons, calcium salt or lignosulfonate is frequently added to them.

Drilling fluids may also be either inhibitive or noninhibitive (Houghton, 1981). The first does not alter the formation once it is cut by the bit. In contrast to the simpler noninhibitive fluids, they inhibit disintegration and retard hydration of drilled solids and commercial (added) clays, and they stabilize the borehole.

Brief descriptions, drawn largely from IMCO Services (1978), indicate the natures and utilities of the eight generic fluids described in Table 7.

1. Potassium/polymer fluids are inhibitive fluids used for drilling through soft formations like shale where sloughing may occur. Polymers are used to maintain their viscosity. These fluids require little thinning with fresh or salt water.

2. Seawater/lignosulfonate fluids are inhibitive fluids that function well under a variety of conditions. They are thought to maintain viscosity by binding lignosulfonate cations onto the broken edges of clay particles, reducing flocculation and maintaining gel strength. They control fluid loss and maintain borehole stability. They are easily altered for more complicated downhole conditions, e.g., higher temperatures.

TABLE 7 Generic Fluid Systems (EPA Region II)

| Type of Fluid | Components | Permissible Content (lb/bbl) |
| --- | --- | --- |
| (1) Potassium/ polymer | Barite | 0-450 |
| | Caustic soda | 0.5-3 |
| | Cellulose polymer | 0.25-5 |
| | Drilled solids | 20-100 |
| | Potassium chloride | 5-50 |
| | Seawater or fresh water | As needed |
| | Starch | 2-12 |
| | XC polymer | 0.25-2 |
| (2) Seawater/ lignosulfonate | Attapulgite or bentonite | 10-50 |
| | Barite | 25-450 |
| | Caustic soda | 1-5 |
| | Cellulose polymer | 0.25-5 |
| | Drilled solids | 20-100 |
| | Lignite | 1-10 |
| | Lignosulfonate | 2-15 |
| | Seawater | As needed |
| | Soda ash/sodium bicarbonate | 0-2 |
| (3) Lime | Barite | 25-180 |
| | Bentonite | 10-50 |
| | Caustic soda | 1-5 |
| | Drilled solids | 20-100 |
| | Fresh water or seawater | As needed |
| | Lignite | 0-10 |
| | Lignosulfonate | 2-15 |
| | Lime | 2-20 |
| | Soda ash/sodium bicarbonate | 0-2 |
| (4) Nondispersed | Acrylic polymer | 0.5-2 |
| | Barite | 25-180 |
| | Bentonite | 5-15 |
| | Drilled solids | 20-70 |
| | Fresh water or seawater | As needed |
| (5) Spud (slugged intermittently with seawater) | Attapulgite or bentonite | 10-50 |
| | Barite | 0-50 |
| | Caustic soda | 0-2 |
| | Lime | 0.5-1 |
| | Seawater | As needed |
| | Soda ash/sodium bicarbonate | 0-2 |

TABLE 7 (continued)

| Type of Fluid | Components | Permissible Content (lb/bbl) |
|---|---|---|
| (6) Seawater/ freshwater gel | Attapulgite or bentonite | 10-50 |
| | Barite | 0-50 |
| | Caustic soda | 0.5-3 |
| | Cellulose polymer | 0-2 |
| | Drilled solids | 20-100 |
| | Lime | 0-2 |
| | Seawater or fresh water | As needed |
| | Soda ash/sodium bicarbonate | 0-2 |
| (7) Lighly treated lignosulfonate freshwater/ seawater | Barite | 0-180 |
| | Bentonite | 10-50 |
| | Caustic soda | 1-3 |
| | Cellulose polymer | 0-2 |
| | Drilled solids | 20-100 |
| | Lignite | 0-4 |
| | Lignosulfonate | 2-6 |
| | Lime | 0-2 |
| | Seawater-to-freshwater ratio | 1:1 approximately |
| (8) Lignosulfonate freshwater | Barite | 0-450 |
| | Bentonite | 10-50 |
| | Caustic soda | 2-5 |
| | Cellulose polymer | 0-2 |
| | Drilled solids | 20-100 |
| | Fresh water | As needed |
| | Lignite | 2-10 |
| | Lignosulfonate | 4-15 |
| | Lime | 0-2 |
| | Soda ash/sodium bicarbonate | 0-2 |

SOURCE: Adapted from Ayers, Sauer, and Anderson (1983).

TABLE 8 Characterizations of Field Drilling Fluids Used in the Joint Industry Mid-Atlantic Bioassay Program

| | General Fluid Types | | | | | | | |
|---|---|---|---|---|---|---|---|---|
| | (1) KCl/ Polymer | (2) SW Ligno- sulfonate[a] | (3) Lime | (4) Nondispersed | (5) SW Spud | (6) SW/FW Gel | (7) LT Ligno- sulfonate | (8) Ligno- sulfonate FW |
| **Components (lb/bbl)** | | | | | | | | |
| Barite | 18.0 | 176 | 64.0 | 10.8[b] | 2 | 21.2 | 9.0 | 15.1 |
| Bentonite/drill solids | 18.0 | 32.1 | 20.0/30.0 | 20.0[b]49.0 | 22.0/52.0 | 9.7[b]14.1[b] | 25.0/48.0 | 15.1/28.1 |
| Chrome lignosuulfonate | 0 | 1.8[b](2.8)[c] | 3.5 | 0 | 0 | 0 | 4.0 | 1.7[b] |
| Lignite | 0 | 0.9[b] | 1.8 | 0.1 | 0 | 0 | 5.0[b] | 2.8[b] |
| Polyanionic cellulose | 1.0 | 0.2[b] | 0 | 1.0 | 0 | 0.5 | 0.5 | 0 |
| Caustic soda | 2.0 | 0.9[b] | 1.5 | [d] | [d] | 0.4[b] | [e] | 1.2[b] |
| Other | KCl (16.0) | (10.0)/ Salt | (1.5)[b] Lime | | | (0.1)/ CMS | | (0.1)/ Lime |
| **Properties** | | | | | | | | |
| Fluid density (lb/gal) | 9.3 | 12.1 | 10.4 | 9.4 | 9.2 | 9.1 | 9.6 | 9.3 |
| Percent solids (wt %) | 18.3 | 43.5 | 27.8 | 21.0 | 21.7 | 11.6 | 24.1 | 16.4 |
| pH | 11.5 | [e] | 10.0 | [e] | [e] | [e] | 10.8 | 9.0 |
| Chlorides (mg/l) | 38,000 | [e] | [e] | 1,200 | [e] | 250 | 7,500 | 1,800 |
| Calcium (mg/l) | [e] | 650 | lime | [e] | [e] | 40 | [e] | 40 |
| Oil and grease[f] | 2,200 | 1,800 | 180 | 290 | 70 | 40 | 50 | 80 |

| Metals[g] (ppm--whole fluid) | | | | | | | | |
|---|---|---|---|---|---|---|---|---|
| Arsenic | 1 | 2 | 3 | 2 | 3 | 2 | 1 | 3 |
| Barium | 24,800 | 141,000 | 76,200 | 13,000 | 2,800 | 25,600 | 11,500 | 14,000 |
| Cadmium | 1 | 1 | 1 | 1 | 1 | 1 | 1 | 1 |
| Chromium | 14 | 227 | 192 | 10 | 16 | 2 | 265 | 48 |
| Copper | 2 | 11 | 8 | 7 | 5 | 2 | 26 | 4 |
| Lead | 2 | 1 | 4 | 2 | 4 | 1 | 24 | 9 |
| Mercury | 1 | 1 | 1 | 1 | 1 | 1 | 1 | 1 |
| Nickel | 6 | 8 | 3 | 4 | 6 | 1 | 6 | 8 |
| Vanadium | 9 | 18 | 27 | 22 | 35 | 6 | 30 | 18 |
| Zinc | 20 | 181 | 58 | 16 | 21 | 12 | 82 | 15 |

[a] Acronyms explained in Table 7.
[b] Estimated concentration outside range designated in generic fluid systems (Table 1).
[c] Chrome lignosulfonate concentration estimate from chromium content calculation (3% Cr in chrome lignosulfonate).
[d] Other components in DCl fluid: soda ash (4.0), aluminum stearate (0.5), sawdust (<.1), lime (<.1), surfactant (<.01), no paraformaldehyde.
[e] Not measured.
[f] Oil and grease analyses conducted by Energy Resources, Cambridge, Mass.
[g] Metals analysis conducted by SCR, Houston, Tex.

SOURCE: Adapted from Ayers, Sauer, and Anderson (1883).

TABLE 9 Summary of Bioassay Results of Mid-Atlantic Generic Drilling Fluids[a]

| Type of Drilling Fluid | 96 hour LC50 in ppm For Mysid Shrimp[a] | | Percent Survival of Hard Shell Clams |
|---|---|---|---|
| | Liquid Phase | Suspended Particulate Phase | Solid Phase (Controls) |
| (1) Potassium/polymer | 66,000[c]<br>58,000[d] | 25,000[b]<br>70,900[d] | 90(99)[c]<br>88(100)[d] |
| (2) Lignosulfonate sea-water | 283,500<br>880,000 | 53,200<br>870,000 | 83(100)[e]<br>70(94)[e] |
| (3) Lime | 393,000<br>1,000,000 | 66,000<br>860,000 | 100(100)<br>94(100) |
| (4) Nondispersed | >1,000,000<br>>1,000,000 | >1,000,000<br>>1,000,000 | 100(100)<br>100(100) |
| (5) Seawater spud | >1,000,000<br>>1,000,000 | >1,000,000<br>>1,000,000 | 100(100)<br>100(100) |
| (6) Seawater/fresh-water gel | >1,000,000<br>>1,000,000 | >1,000,000<br>>1,000,000 | 100(100)<br>100(100) |
| (7) Lightly treated lignosulfonate fresh water/sea-water | >1,000,000<br>>1,000,000 | >1,000,000<br>>1,000,000 | 97(98)<br>100(100) |
| (8) Lignosulfonate fresh water | >1,000,000<br>>1,000,000 | 506,000<br>>1,000,000 | 99(100)<br>99(100) |

NOTE: Characterization of Toxicity (IMCO/FAO/UNESCO/WMO, 1969)

| LC50 Value (ppm) | Toxicant Classification |
| --- | --- |
| >10,000 | Practically nontoxic |
| 1,000-10,000 | Slightly toxic |
| 100-1,000 | Moderately toxic |
| 1-100 | Toxic |
| <1 | Very toxic |

a LC50 values are expressed as ppm and must be multiplied by 0.20 to obtained values for drilling fluid used to formulate phases.

b Physical phases of drilling fluids were extracted from a 1:4 mixture by volume of fluid and synthetic or natural sea water. Test organism for the liquid and suspended particulate phases was the mysid shripm (*Mysidopsis bahia*), and for the solid phase was the hard shell clam (*Mercenaria mercenaria*). Protocol for testing was established by EPA Region II in conjunction with the Mid-Atlantic Operators.

c First values given in these columns were determined by Energy Resources, Cambridge, Massachusetts.

d Second values given in these columns were determined by Normandeau Associates, Bedford, New Hampshire.

e Statistically significant differences ($\alpha = 0.05$) in survival between clams exposed to the solids phase of fluid and control sediment.

3. Lime (or calcium) fluids are inhibitive fluids in which calcium binds onto clay. The clay platelets are pulled together, dehydrating them and releasing absorbed water. The size of the particles is reduced, and water is released, resulting in reduced viscosity. More solids may be maintained in these systems with a minimum of viscosity and gel strength. These fluids are used in hydratable, sloughing shale formations.

4. Nondispersed fluids are inhibitive fluids in which acrylic serves to prevent fluid loss and maintain viscosity. They also provide improved penetration, which is impeded by clay particles in dispersed fluids.

5. Spud fluids are noninhibitive, simple mixtures used in the first 1,000 (300m) or so of drilling.

6. Seawater/freshwater gel fluids are inhibitive fluids used early in drilling or in simple drilling situations. They provide good fluid control, shear thinning, and lifting capacity. Prehydrated bentonite that flocculates is used in such freshwater or saltwater fluids. Attapulgite is used in saltwater fluids when fluid loss is not important.

7. Lightly treated lignosulfonate freshwater/seawater fluids resemble seawater/lignosulfonate fluids (type 2) except that the salt content is less. The viscosity and gel strength of these fluids are adjusted through additions of lignosulfonate and caustic soda.

8. Lignosulfonate freshwater fluids resemble fluid types 2 and 7, except that lignosulfonate concentrations are higher. These fluids are suited to high-temperature drilling. Increased concentrations of lignosulfonate will result in heavily treated fluids of this type.

As the descriptions of the generic fluids indicate, these fluids share numerous properties. The major ones are containing either fresh water or seawater, being inhibitive or noninhibitive, and being non-dispersed or lignosulfonate-treated polymers. Certain components are shared by fluids in each of these categories, for example, the weighting agent barite, and the caustic soda used to control pH.

The concept of generic drilling fluids was developed initially for exploratory wells. Its application to development wells (the majority of those drilled) is recent. The differences between discharges from exploratory and development wells have been assessed (Boothe and Presley, 1983), and can be addressed within the framework of generic drilling fluids.

## THE REGULATION OF DRILLING DISCHARGES

Principal authority to regulate the discharge of drilling fluids and cuttings in offshore oil and gas activities rests with EPA through its National Pollutant Discharge Elimination System (NPDES), which was established under Section 402 of the Clean Water Act (formerly the Federal Water Pollution Control Act Amendments). The Minerals Management Service (MMS) of the Department of the Interior also controls discharges through lease stipulations and OCS operating orders under the authority of the Outer Continental Shelf Lands Act Amendments of 1978.[9]

The Clean Water Act requires that point-source discharges of pollutants achieve effluent limitations through use of the "best practicable control technology currently available" (BPT). EPA determines BPT limitations for categories of industrial discharges and promulgates national guidelines for regional permits concerning the pollution control a discharger will achieve while utilizing BPT.

BPT limitations relevant to drilling fluids are contained in the limitations for the oil and gas extraction industry (40 CFR 435). Current limitations mention only oil and grease. These adopt the "no free oil standard" established under the oil discharge liability provision of Section 311 of the Clean Water Act. This standard prohibits any discharge that would cause a film or sheen on the surface of the water or a sludge or emulsion to be deposited beneath the surface of the water (40 CFR 110). Discharges that cannot meet this standard are to be disposed of on land at a dump site approved under RCRA.

Under Sec. 301(c) of the Clean Water Act, EPA is currently developing standards concerning the Best Available Technology Economically Achievable (BAT). These standards may include a prohibition on the use of diesel fuel, requirements for bioassays, the use of generic fluid categories, and new compliance tests.

Section 306 of the Clean Water Act requires new source performance standards for discharges through application of the "best available demonstrated control technology", reflecting the greatest degree of effluent reduction. Such standards have yet to be promulgated for drilling fluids.

NPDES permits, which are issued through EPA's regional offices, must be preceded by determinations under Section 403(c) of the Clean Water Act that the discharges will not result in unreasonable degradation of the marine environment. This section, and its implementing regulations, the Ocean Discharge Criteria (40 CFR Part 125) issued in 1980,[10] provide a two-tiered test of degradation. Based on information supplied by the applicant and other relevant material, the regional administrator assesses the potential for "unreasonable degradation": significant adverse changes in ecosystem diversity and productivity and in the stability of the biological communities within and surrounding the area of discharge; threats to human health through direct exposure to pollutants or consumption of exposed aquatic organisms; or loss of aesthetic, recreational, scientific or

---

[9] In territorial waters, states may also impose requirements on discharges, either through administration of NPDES permit programs (where states have been delegated such authority by EPA), or through separate state regulations. States cannot be less restrictive than the EPA in administering NPDES permit programs.

[10] Prior to 1980, the issuing of permits was guided by the ocean dumping regulations, 40 CFR 227, which require bioassays and the calculation of the "limiting permissible concentration" (LPC) of the discharge following dilution.

economic values unreasonable in relation to the benefits derived from the activities leading to the discharge. Such determinations depend on the location of the discharge, the presence of special aquatic sites, and the nature of the discharge, including its composition, potential toxicity through bioaccumulation, and persistence and transport in the marine environment.

If the proposed discharge is found not likely to cause unreasonable degradation, then it may be permitted. If information is insufficient to determine whether unreasonable degradation will occur, no permit may be issued unless another determination is made that the discharge will not cause "irreparable harm." Irreparable harm is defined as significant undesirable effects, occurring after permit issuance, that will not be reversed by ceasing or modifying the discharge (Section 125.121(a) of the Clean Water Act). In such cases, it must be judged that the discharge will not result in irreparable harm during the period in which monitoring can be conducted, and that there are no reasonable alternatives to onsite disposal of the wastes. The discharge must meet a number of conditions, among them: it may not exceed a limiting permissible concentration (LPC) for the liquid and suspended particulate phases of the waste following dilutions measured from the boundary of a mixing zone(defined as 100 m from the point of discharge); it may not exceed the LPC for the solid phase or result in bioaccumulation; permit conditions may require environmental monitoring of discharges or other appropriate conditions.

Drilling fluids determined unacceptable for disposal under the Ocean Discharge Criteria or under State authority in territorial waters may be considered for ocean dumping at a designated ocean dump site for land disposal. In federal notes of discharges and dumpsite designations are authorized under Title I (The "Ocean Dumping Act") of the Marine Protection, Research and Sanctuaries Act. The regulations that implement this Act (40 CFR 220-229) provide for the calculation of a limiting permissible concentration based on liquid, suspended particulate, and solid phase bioassays. Land disposal is regulated under the Reseource Conservation and Recover, Act (RCRA).

EPA's Region IX (San Francisco) issued the first offshore NPDES permit, to the Shell Oil Company, in 1976. It later issued permits for the Atlantic and the Gulf of Mexico. Also, wells have been drilled offshore Alaska with EPA concurrence. It developed a general permit now in force in the Gulf of Mexico and California but also issues in these regions individual permits that are designed to protect biologically sensitive areas (e.g., the Flower Garden Banks in the Gulf of Mexico).

In addition to, or as adjuncts to, the Ocean Discharge Criteria and the effluent limitations, NPDES permits may make special prohibitions (e.g., on the use of pentachlorophenol or asbestos), require special discharge practices (e.g., shunting to the nepheloid layer or predilution), and require biological or other studies to monitor the marine environment for changes as the result of discharges. These conditions may complement those imposed by MMS. For example, EPA and MMS jointly required and aided in developing a biological monitoring program for Georges Bank.

Before EPA exercised its authority over offshore drilling discharges, the Bureau of Land Management and the Conservation Division of the U.S. Geological Survey (now combined as MMS) placed special requirements on operators through lease stipulations and OCS operating orders (Rieser and Spiller, 1980). Lease stipulations commonly give the MMS district supervisor the authority to require special discharge practices when appropriate, for example, district supervisors may specify monitoring programs and depths at which discharges are to be released in biologically sensitive areas. The objective of operating orders is to ensure safe operations. MMS Operating Order 7 specifically addresses pollution, and, while noting that fluid disposal is subject to the requirements of EPA, this order also requires information on the constitutents of drilling fluids and additives. The use offshore of pentachlorophenol is prohibited under this MMS authority.

The NPDES permits of the Environmental Protection Agency and the operating orders and other requirements of the Minerals Management Service (and consequently industrial operating practices, including the use of additives) vary according to geographic regions. For example, in 72 percent of wells in a sample in the Gulf of Mexico in 1982, additives were used that were not approved for use in EPA Region II (a mid-Atlantic region), where the concept of generic drilling fluids has been adopted (Dalton, Dalton, Newport, 1983). Regional differences have been taken into account because of environmental conditions, or to protect sensitive or valuable habitats.[11] (See Table 23.)

Government regulation has spurred extensive research, both in anticipation of permit conditions and as a result of those conditions. EPA's initial attempts to regulate drilling discharges were repeatedly challenged by industry. More recently, however, there has been growing cooperation between industry and government, resulting in the development of monitoring programs (e.g., the Georges Bank monitoring program), the Region II bioassay protocol (U.S. Environmental Protection Agency Region II, 1978), a program for sampling used drilling fluids ("PESA muds") and for conducting toxicity testing and chemical analyses of them, and the specification of generic drilling fluids (Ayers, Sauer, and Anderson, 1983).

## THE MASS LOADING OF DRILLING DISCHARGES IN RELATION TO THAT OF OTHER INPUTS TO THE MARINE ENVIRONMENT

As part of the review of the discharge of drilling fluids and cuttings, the quantities and frequencies of these discharges and their components will be examined in light of their mass loading into the coastal ocean compared to that of sediments and trace metals from other natural and anthropogenic sources. Because these inputs vary greatly with time and place, the comparisons that follow do not necessarily reflect relative effects. They are useful, however, in considering the magnitude of

---

[11] An alternative explanation for the difference in regulations for the Gulf of Mexico and EPA Region II is that the list of approved additives for EPA Region II was not up to date in 1982 because there had been no drilling there since 1981.

drilling-fluid discharges on the OCS. Caution is advised concerning the estimates used--most are approximations.

## Drilling Discharges

Several estimates of the magnitude of drilling-fluid discharges on the OCS are available (see Table 10). These result from theoretical calculations or product use inventories from exploratory, development, and production wells in the Gulf of Mexico, mid-Atlantic, and North Atlantic (Georges Bank) OCS areas. These statistics display approximately a fourfold range in mean total solids and chromium discharged per well, but a much narrower range in mean barium discharge. Most of the variation is due to the fact that smaller discharges are made from the shallower development and production wells in the Gulf of Mexico.

## River Inputs

Table 11 compares the average annual discharges of water and sediments of major rivers in North America. Although sediment discharge is generally related to drainage area and water discharge, some rivers (e.g., the Eel River in northern California and the Copper River in Alaska) contribute disproportionately large loads of sediment to the sea. Sediment discharges are frequently episodic and highly variable from year to year. For example, in 1969 the Santa Clara River in California flooded and discharged $1 \times 10^8$ metric tons (t) of sediment during 2 weeks, compared to an annual average discharge of $2 \times 10^6$ t. This event increased sedimentation in the Santa Barbara Basin 10 to 100 mm compared with the long-term annual sedimentation rates of 1 to 5 mm.

The total loading in 1980 of particulate material from drilling discharges on the U.S. OCS is estimated as $1.85 \times 10^6$ t, compared to over $4 \times 10^8$ t per year for North American rivers. Most of the load from drilling discharges was in the northern Gulf of Mexico, and was equivalent to approximately 0.8 percent of the Mississippi River's input to the Gulf.

Barium, added as barite ($BaSO_4$), is commonly present in drilling fluids at much higher concentrations than in marine or riverine sediments and thus serves as an effective tracer of drilling-fluid contamination of marine sediments. Barium is present in an average concentration of 62 μg/l in Mississippi River water (Hanor and Chan, 1977), but most of the barium discharged by rivers is in relatively insoluble particulate material in an average concentration of 600 μg/g (dry weight) of suspended sediment (Martin and Maybeck, 1979).

The average mass emission of barium by the Mississippi River is approximately $1.5 \times 10^5$ t per year, almost all of which is particulate (Table 12). The release of barium from OCS drilling activities has been estimated to be $3.2 \times 10^5$ t per year. Since these estimates are approximations, it is perhaps more appropriate to say that the mass emissions of barium from OCS drilling discharges appear to be of the same order of magnitude as those from the Mississippi River.

TABLE 10 Average Discharges of Particulate Solids, Barium, and Chromium from OCS Wells

| | Gulf of Mexico, Average Well (Gianessi and Arnold, 1982) | Gulf of Mexico, Five Exploratory Wells (Petrazzuolo, 1981) | Mid-Atlantic, Exploration Well (Ayers et al., 1980) | Georges Bank, Eight Exploratory Wells (Danenberger, 1983) | Gulf of Mexico, Forty-Nine Exploration, Development and Production Wells (Boothe and Presley, 1983) |
|---|---|---|---|---|---|
| Depth of well (m) | 5,486 | 3,329 | 4,970 | 4,900 | 3,121 |
| Total solids (t) | 1,140 | | 2,160 | 1,220[a] | 598 |
| Barite | - | 600 | 752 | 715 | 492 |
| Barium [b] (t) | - | 312 | 391 | 372 | 256 |
| Chrome lignosulfonate | - | 20 | 45 | 26 | 10 |
| Chromium [c] (t) | - | 0.6 | 1.3 | 0.8 | 0.3 |

[a] Drilling fluid solids only (does not include cuttings).
[b] Barium estimated at 52 percent of barite weight.
[c] Chromium estimated at 2.9 percent of chrome lignosulfonate weight.

TABLE 11 Drainage Area and Water and Suspended Sediment Discharges of North America's Major Rivers

| River | Drainage Area (millions of $km^2$) | Water Discharge ($km^3$ per year) | Sediment Discharge (t per year) |
|---|---|---|---|
| St. Lawrence (Canada) | 1.03 | 447 | 4 |
| Hudson (USA) | 0.02 | 12 | 1 |
| Mississippi (USA) | 3.27 | 580 | 191 |
| Brazos (USA) | 0.11 | 7 | 15 |
| Colorado (Mexico) | 0.64 | 20 | 0.1 |
| Eel (USA) | 0.008 | - | 13 |
| Columbia (USA) | 0.67 | 251 | 7 |
| Fraser (Canada) | 0.22 | 112 | 18 |
| Yukon (USA) | 0.84 | 195 | 55 |
| Copper (USA) | 0.06 | 39 | 64 |
| Susitna (USA) | 0.05 | 40 | 23 |
| Mackenzie (Canada) | 1.81 | 306 | 91 |

SOURCE: Adapted from Milliman and Meade (1983).

Chromium, added mainly in the form of chrome lignosulfonate, may also be more concentrated in drilling fluids compared to riverine sources (Table 12). Mass emissions of chromium associated with suspended sediments in North American rivers are estimated as $76 \times 10^3$ t per year, while the dissolved input has been estimated at $22 \times 10^3$ t per year. Emissions of chromium from drilling operations (estimated from usage of additives or analysis of discharged material) averages about 580 μg/g dry weight of solids. In contrast to barium, however, much of the chromium in drilling fluids is soluble and will disperse differently from particulate components when discharged into the ocean. Drilling discharges of chromium equal just over 1 percent of the input of North American rivers.

## Anthropogenic Wastes

A broad variety of other wastes, including municipal sewage, industrial wastes, and dredged material, is introduced into both coastal and OCS waters via pipelines, barges, ships, and offshore drilling vessels and platforms. Table 13 compares direct waste inputs into the U.S. coastal ocean, including drilling-fluid discharges. It should be kept in mind that the concentrations, bioavailabilities, and geographic locations of such inputs vary greatly and consequently so do their effects. Thus, Table 13 does not compare the environmental significance of these wastes. It does indicate that the mass emissions of dredged materials, sewage sludge, and industrial wastes exceed those of drilling fluids.

The amount of suspended solids in southern California municipal waste discharges is approximately $2.5 \times 10^5$ t per year (Bascom, 1982), that in drilling-fluid discharges on the California OCS is $1.7 \times 10^4$ t per year. These municipal wastes include approximately 230 t per year of chromium; the figure for California drilling discharges is roughly 10 t per year. The introduction of chromium from U.S. OCS drilling discharges, approximately $1 \times 10^3$ t per year (Table 12), approaches that from waste disposal in the New York Bight, $1.4 \times 10^3$ t per year (Mueller et al., 1976).

## Other Human Impacts

Discharged drilling fluids and cuttings may settle on the bottom and harm benthic organisms within some area around the rig. These effects may be primarily physical and, providing that bottom sediments are not modified over a long period of time, may disturb the seabed much in the way that storms, dredging, the disposal of dredged material, and certain fishing activities do.

Dredging for surf clams *Spisula solidissma* covers average swathes 1.5 m wide by 46 cm deep (Ropes, 1972), which might disturb $4.3 \times 10^3$ $m^3$ of sediment per vessel per day. There were 98 surf clam boats working along the U.S. east coast in 1974 (Ropes, 1982). In contrast, Gianessi and Arnold (1982) estimated that an average of 442 $m^3$ of drill solids are discharged per well over approximately 90 days. (Regarding this comparison, it needs to be kept in mind that fishing

TABLE 12 Estimates of Mass Emissions of Particulate Solids, Barium, and Chromium from OCS Drilling Discharges and from Rivers

| Source | Sediment Discharged (t per year) | Barium Concentration (mg/g) | Barium Loading (t per year) | Chromium Concentration (μg/g) | Chromium Concentration (t per year) |
|---|---|---|---|---|---|
| U.S. OCS Drilling Fluids | $1.3 \times 10^6$ | 250[a] | $3.2 \times 10^5$[a] | 580 | $1.1 \times 10^3$ |
| North American Rivers | $4 \times 10^8$[b] | | | 190[c] | $76 \times 10^3$ |
| Mississippi River | $2.1 \times 10^8$[b] | 0.74[d] | $1.5 \times 10^5$ | 150[c] | $31 \times 10^3$ |

[a]Estimated from the average barium discharge per OCS well inventoried (Table 10) and 1,033 wells drilled in 1981. It is not known how representative these wells are with regard to barium concentration.
[b]Milliman and Meade (1983).
[c]Goldberg (1980).
[d]Trefry et al. (1981).

TABLE 13 Ocean Disposal of Various Wastes by Geographic Areas, 1973 to 1980 (Millions of Tons)

| | Atlantic Ocean | | | Gulf of Mexico | | | Pacific Ocean | | | Total | | |
|---|---|---|---|---|---|---|---|---|---|---|---|---|
| | 1973 | 1976 | 1980 | 1973 | 1976 | 1980 | 1973 | 1976 | 1980 | 1973 | 1976 | 1980 |
| Industrial Wastes[a] | 3.643 | 2.633 | 2.928 | 1.408 | 0.100 | 0 | 0 | 0 | 0 | 5.051 | 2.733 | 2.928 |
| Sewage Sludge[a] | 4.898 | 5.271 | 7.309 | 0 | 0 | 0 | 0 | 0 | 0 | 4.898 | 5.271 | 7.309 |
| Construction and Demolition Debris[a] | 0.974 | 0.315 | .089 | 0 | 0 | 0 | 0 | 0 | 0 | 0.974 | 0.315 | 0.089 |
| Dredged Wastes[b] | - | - | - | - | - | - | - | - | - | 89.376 | 92.485 | 60.866 |
| Wood Incineration[a] | 0.011 | 0.009 | 0.011 | 0 | 0 | 0 | 0 | 0 | 0 | 0.011 | 0.009 | 0.011 |
| Drilling Fluids[c] and Cuttings | 0 | 0 | 0.008 | 1.94 | 1.85 | 1.67 | 0.12 | 0.082 | 0.17 | 2.06 (2.01%) | 1.932 (1.88%) | 1.848 (2.5%) |
| | | | | | | | | | | 102.87 | 102.695 | 73.051 |

[a] Source: EPA (1980).
[b] Source: U.S. Army Corps of Engineers (1982). Original values of million cubic yards of material converted to tons by applying multiplier of 1.33 tons/yd$^3$. This value can vary significantly depending on material. This multiplier was delivered from 1980 data given in this source.
[c] Source: Quantities based on estimated value of 2,000 t per well (dry weight) of fluids and cuttings solids. Pacific includes wells drilled off Alaska.

dredges and trawls disturb <u>in situ</u> sediment, with attendant physical effects; the discharge of drilling fluids and cuttings adds foreign substances to the marine environment.)

These comparisons to river inputs, anthropogenic wastes and other human impacts are made not to suggest that the effects of drilling discharges are minimal by comparing them to traditional uses of the ocean's natural resources; rather they indicate that the use of natural resources virtually always results in some potentially undesirable side effects. With each activity, appropriate and effective pollution prevention and mitigation measures are needed. Regardless of the relative contributions of pollutants to the marine environment from other sources, it is the mandate of this study to provide an effective assessment of the environmental risks of drilling fluids and cuttings. In meeting that charge, the chapters of this report provide more detailed consideration of the compositions, locations, and frequencies of drilling discharges; their fates, including dispersion and chemical transformation; and their effects on marine biota.

## REFERENCES

American Petroleum Institute. 1978. Oil and gas well drilling fluid chemicals. 1st ed. API Bull. 13F.

American Petroleum Institute. 1982. Total number of offshore wells drilled in the United States. In: Basic Petroleum Data Book. Washington, D.C.: American Petroleum Institute. Vol. 3, No. 1, Sec. H, Table 7.

Ashford, N.A., and G.R. Heaton. 1979. The effects of health and environmental regulation on technological change in the chemical industry: theory and evidence. In: C.T. Hill (ed.), Federal Regulation and Chemical Innovation. American Chemical Society Symposium Series 109, September 14, 1978, Miami Beach, Fla. Washington, D.C.: American Chemical Society. Xv + 200 pp. Pp. 45-66.

Ayers, R.C., Jr. 1983. Characteristics of drilling discharges. In: Proceedings of Workshop on Modeling, Santa Barbara, Calif. Reston, Va.: Minerals Management Service. In press.

Ayers, R.C., Jr. 1981. Fate and effects of drilling discharges in the marine environment. Proposed North Atlantic OCS oil and gas lease sale 52. Statement delivered at public hearing, Boston, Mass., November 19, 1981. Bureau of Land Management, U.S. Department of the Interior.

Ayers, R.C., Jr., T.C. Sauer, Jr., and P. Anderson. 1983. The generic mud concept for offshore drilling for NPDES. Paper presented at the IADC/SPE 1983 Drilling Conference. Paper No. IADC/SPE 11399. Available from SPE, 6200 North Central Causeway, Drawer 64706, Dallas, TX 75206.

Ayers, R.C., Jr., T.C. Sauer, Jr., R.P. Meek, and G. Bowers. 1980. An environmental study to assess the impact of drilling discharges in the mid-Atlantic. In: Proceedings of a Symposium on Research on Environmental Fate and Effects of Drilling Fluids and Cuttings. Washington, D.C.: Courtesy Associates. P. 382.

Bascom, W. 1982. The effects of waste disposal on the coastal waters of Southern California. Environ. Sci. Techol. 1664:226A-236A.

Boothe, P.N., and B.J. Presley. 1983. Distribution and behavior of drilling fluid and cuttings around Gulf of Mexico drill sites. Draft final report. API Project No. 243. Washington, D.C.: American Petroleum Institute.

Carney, L.L., and L. Harris. 1975. Thermal degradation of drilling mud additives. In: Proceedings, Environmental Aspects of Chemical Use in Well-Drilling Operations, May 21-23, 1975, Houston, Tex. EPA-5601/1-75-004. Washington, D.C.: Office of Toxic Substances, U.S. Environmental Protection Agency. vi + 604 pp.

Dalton, Dalton, Newport. 1983. Analysis of drilling muds from 74 offshore oil and gas wells in the Gulf of Mexico. Draft report dated May 6, 1983. Available from Monitoring and Data Support Division, Office of Water Regulations and Standards, U.S. Environmental Protection Agency, Washington, D.C.

Danenberger, Elmer P. 1983. Georges Bank exploratory drilling 1981-1982. n.p. Available from Minerals Management Service, Offshore Division, Reston, Va.

Dolton, G.L., et al. 1981. Estimates of undiscovered recoverable resources of conventionally producible oil and gas in the United States, a summary. Open File Report 81-192. Reston, Va.: U.S. Geological Survey. 17 pp.

Edgar, N. Terrance. 1983. Oil and gas resources of the U.S. continental margin. In: Gerald J. Mangone (ed.), The Future of Oil and Gas from the Sea. New York: Van Nostrand Reinhold Co., Inc. Pp. 25-79.

Escott, T.A., and J.B. Walker. 1981. The drilling fluids market: status report. Oil Service Monthly Bulletin, June 1981. Paine, Webber, Mitchell, Hutchins, Inc. 11 pp.

Gianessi, L.P., and F.D. Arnold. 1982. The discharges of water pollutants from oil and gas exploration and production activities in the Gulf of Mexico region. Draft report prepared by Resources for the Future, Washington, D.C. Contract NA-80-SAC-00793. Office of Ocean Resources Coordination and Assessment, NOAA.

Goldberg, E.D. (ed). 1980. Proceedings of a Workshop on Assimilative Capacity of U.S. Coastal Waters for Pollutants,

Crystal Mountain, Washington, July 29-August 4, 1979. Working Paper No. 1: Federal Plan for Ocean Pollution Research Development and Monitoring, FY 1971-1985. Second Printing--Revised. Environmental Research Laboratory, NOAA, U.S. Department of Commerce, Boulder, Colo.

Gray, G.R., H.C.H. Darley, and W.F. Rogers. 1980. Composition and Properties of Oil Well Drilling Fluids. 4th ed. Houston, Tex.: Gulf Publishing Co. xii + 630 pp.

Hanor, J.S., and H.L. Chan. 1977. Non-conservative behavior of barium during mixing of Mississippi River and Gulf of Mexico water. Earth Planet. Sci. Lett. 37:242-250.

Houghton, J.P., et al. 1981. Fate and effects of drilling fluids and cuttings discharges in lower Cook Inlet, Alaska, and on Georges Bank. Report prepared for Branch of Environmental Studies, Minerals Management Service. Dames & Moore, Inc., Seattle, Wash.

IMCO Services. 1978. Applied Mud Technology. Houston, Tex.: IMCO Services, Halliburton Co.

Intergovernmental Maritime Consultative Organization, et al. 1969. Abstract of first session report of Joint Group of Experts on the Scientific Aspects of Marine Pollution. Water Res. 3:995-1005.

Jones, M., C. Collins, and D. Havis. 1980. Trade-offs in traditional criteria vs. environmental acceptability in product development: an example of the drilling fluids industry's response to environmental regulations. Environ. Prof. 2:94-102.

Jones, M., and M. Hulse. 1982. Drilling fluid bioassays and the OCS. Oil Gas J. 80(25):241-244.

McGuire, E.P. 1973. Evaluating new product proposals. Conference Board Report No. 604. New York: The Conference Board.

Martin, J., and M. Maybeck. 1979. Elemental mass balance of material carried by major world rivers. Mar. Chem. 7:173-206.

Milliman, J.D., and R.H. Meade. 1983. Worldwide delivery of river sediments to the oceans. J. Geol. 91:1-21.

Minerals Management Service. 1982. Approval of five-year OCS oil and gas leasing program announced. News Release, July 21, 1982. Washington, D.C., Department of the Interior. 7 pp.

Moseley, H.R., Jr. 1981. Chemical components, functions, and uses of drilling fluids. In: Proceedings of UNEP Conference, Paris, France, June 2-4, 1981. P. 43.

McAtee, J.L., and N.R. Smith. 1969. Ferrochrome lignosulfonates. J. Colloid Interface Sci. 29(3):389-398.

Mueller, J.A., J.S. Jeris, A.R. Anderson, and C.F. Hughes. 1976. Contaminant inputs to the New York Bight. Technical Memorandum ERL MESA-6. Rockville, Md.: NOAA. 347 pp.

Palmer, C.R., and P.L. Kelley. 1983. America's five-year offshore leasing plan--its importance in increasing domestic petroleum reserves. Presentation to 23d Annual Institute in Petroleum Exploration.

Perricone, C. 1980. Major drilling fluid additives. In: Proceedings of a Symposium on Research on Environmental Fate and Effects of Drilling Fluids and Cuttings. Washington, D.C.: Courtesy Associates. P. 15.

Petrazzuolo, G. 1981. Preliminary report. An environmental assessment of drilling fluids and cuttings released onto the outer continental shelf. Vol. 1: Technical assessments. Vol 2: Tables, figures and Appendix A. Prepared for Industrial Permits Branch, Office of Water Enforcement and Ocean Programs Branch, Office of Water and Waste Management, U.S. Environmental Protection Agency, Washington, D.C.

Petroleum Extension Service. 1979. A Dictionary of Petroleum Terms. Austin, Tex.: University of Texas Petroleum Extension Service. 129 pp.

Rieser, A., and J. Spiller. 1981. Regulating Drilling Effluents on Georges Bank and the Mid-Atlantic Outer Continental Shelf: A Scientific and Legal Analysis. Boston, Mass.: New England States/ New England River Basins Commissions. 130 pp.

Ropes, J.W. 1972. The Atlantic coast surf fishery--1965-1969. Mar. Fish. Rev. 34(7-8):20-29.

Ropes, J.W. 1982. The Atlantic coast surf fishery--1965-1974. Mar. Fish. Rev. 44(7-8):1-4.

Skelly, W., and D.E. Dieball. 1970. Behavior of chromate in drilling fluids containing chrome lignosulfonate. J. Soc. Pet. Eng.: 140-144.

Skelly, W., and J.A. Kjellstrand. 1966. Thermal degradation of modified lignosulfonates in drilling mud. In: Proceedings, API Production Department, Spring Meeting, March 2-4, 1966, Houston, Tex. Dallas, Tex.: American Petroleum Institute. 12 pp.

Trefry, J.H., R.P. Trocine, and D.B. Meyer. 1981. Tracing the fate of petroleum drilling fluids in the northwest Gulf of Mexico. Oceans '81. Available from Marine Technology Society, Washington, D.C. Pp. 732-736.

U.S. Army Corps of Engineers. May 1982. Report to Congress on administration of ocean dumping activities. Pamphlet 82-P1.

U.S. Environmental Protection Agency. 1978. Bioassay procedures for the Ocean Disposal Permit Program. Report No. EPA-600/9-78-010. U.S. Environmental Protection Agency, Washington, D.C.

U.S. Environmental Protection Agency. 1980. On administration of the Marine Protection, Research, and Sanctuaries Act of 1972. Annual report to Congress.

Van Olphen, H. 1977. An Introduction to Clay Colloid Chemistry. 2d ed. New York: Wiley & Sons. xvii + 318 pp.

Wright, T.R., Jr., and R.D. Dudley (eds.). 1982. Guide to Drilling Completion and Workover Fluids. World Oil. Houston, Tex.: Gulf Publishing Co.

# 3
# The Fates of Drilling Discharges

## INTRODUCTION

The fates of drilling fluids and cuttings discharged in the marine environment are determined by diverse physical processes (current, gravity), chemical processes (reaction, sorption), and biological processes that all serve to disperse or concentrate constituent materials. The various dissolved and particulate constituents behave in different ways when encountering seawater and transport forces in the ocean. Even so, some generalizations can be made, and they allow predicting the fates of drilling fluids.

Although the ocean is a continuous liquid with a long time scale for mixing, it remains an inhomogeneous solution. Inhomogeneities are caused by energetics that set up horizontal density gradients of liquid (fronts) and vertical ones (pycnoclines) through which transfer is relatively slow. Within the boundaries established by density gradients, inhomogeneities tend to be less as a consequence of the conservative nature of the major components of seawater; yet heavy metals, nutrients, dissolved gases and organic matter may be non-conservative both in quantity and chemical form as a result of geochemical and biological processes. The forces of the ocean, however, are continuously at play, reducing these gradients and producing more constant composition. While molecular diffusion[1] in any liquid brings about homogeneity, the rate is slow ($10^{-5}$ $cm^2/s$) compared to the rates

---

[1] Molecular diffusion is the gradual mixing of molecules of two or more substances through random thermal motion. In a solution in which the concentration of a substance varies in space, the amount of that substance which per second diffuses through a surface area of 1 $cm^2$ is proportional to the change in concentration per cm along a line normal to that surface ($dM/dt = \delta\, de/dn$). The proportionality constant ($\delta$) is the diffusion coefficient, which for seawater is about $2 \times 10^{-5}$.

produced by mixing or eddy diffusivity[2] in the ocean (approximately 1 $cm^2/s$ in the vertical and $10^5$ to $10^7$ $cm^2/s$ in the horizontal) (Okubo, 1971).

The ocean's properties and composition are nonuniform because of many factors. Major among these are heat exchange with the atmosphere, which leads to evaporation and surface cooling, which both increase density; and surface warming, river runoff, and rainfall, which all decrease density. Changes in density cause mixing both vertically and horizontally, and, together with winds and tides, provide most of the energy for ocean mixing. Other factors that are not very important to physical mixing also cause inhomogeneities. Phytoplankton growth decreases concentrations of nutrient elements and alters the carbon dioxide components, causing an increase in pH. River runoff adds sediment and dissolved materials, including anthropogenic components derived from various uses of water. Sorption processes, in which trace metals and organic compounds selectively adhere to or exchange on surfaces, also result in the inhomogeneous distribution of materials. For example, trace metals may adsorb to clay minerals, which then are deposited on the seafloor, while other materials, such as polychlorinated biphenyls (PCBs), may concentrate at the sea surface. In addition, fine particulate solids and the associated sorbed materials in suspension often flocculate when mixed with seawater, thereby increasing the settling rate of the solids and altering the physical and chemical characteristics of deposited sediment.

When discharged into the ocean, a material composed of finely divided insoluble materials or solutes immediately is subject to a process of dilution in a concentration gradient decreasing from the point of discharge. To reverse this process (for instance, by bioaccumulation) requires sufficient energy to overcome the dilution process. Components that are held in solution or suspension are rapidly diluted by a factor of $10^5$ to $10^6$ within the first hour (Sverdrup et al., 1942) from the eddy diffusion resulting from the ocean turbulence generated mainly by geostrophic flow[3], tidal currents, and wind mixing (Hill, 1962). Neutrally buoyant or dissolved materials will form a dispersion (dilution) plume, riding the path of currents through the ocean, always decreasing in concentration. Those

---

[2]The rate of transfer of mass in water is proportional to the gradient of concentration. The proportionality coefficient is called eddy diffusivity. It is not a physical constant, but depends on the nature of the turbulent motion. (This motion is that of a liquid having local velocities and pressures that fluctuate randomly. It is also called turbulent flow.) The ranges of eddy diffusivity per unit mass are about 0.1 to 100 $cm^2/s$ in the vertical and about $10^6$ to $10^8$ $cm^2/s$ in the horizontal direction (Sverdrup et al., 1942).

[3]The oceanic flow resulting from the earth's rotation.

materials that are negatively buoyant separate from the suspended plume according to their specific settling characteristics. They may then be reconcentrated by gravity on the seafloor, where they may be buried by physical and biological processes, resuspended and transported, or chemically altered by benthic processes.

The fates of materials discharged into the marine environment are influenced heavily by the dispersive and transport energy of the ocean at the discharge site. This energy dominates the rate of dispersion after the dynamic energy induced by the actual discharge has decayed. In ocean discharge operations this dynamic energy has been used extensively to cause rapid dilution. Discharges in the wakes of moving barges (Hood et al., 1958; Ketchum and Ford, 1952), outfall diffusers (Colonell, 1981; Yudelson, 1967) and high-pressure jets (Brandsma et al., 1980) are very effective ways to reach dilutions of several orders of magnitude within only a few meters of the place of discharge.

Dispersion from a point source into the marine environment varies with site location and depth of discharge because of the variability of several important factors influencing the turbulence (eddy diffusivity) of the water column and the bottom boundary layer:

- Vertical or horizontal stratification by temperature, salinity, and suspended sediments
- Wind and tidal energy interaction
- The topography of large-scale bed forms
- Variable bed conditions (bioturbation, bed forms, and near-bed transport).

These factors vary not only from site to site; storm events also affect ambient flow conditions. These factors together provide a general framework for analyzing the fates of drilling fluids and cuttings.

## BEHAVIOR OF THE DISCHARGE PLUME

The phenomena observed during drilling-fluid discharges are explained by the Offshore Operator's Committee model (Brandsma et al., 1980) which is illustrated in Figure 2. The initial plume is denser than seawater and goes through a stage of convective descent until it encounters the seabed or becomes neutrally buoyant from loss of solids and water entrainment. As a result of the density gradient of the plume, the plume then collapses and goes into a stage of passive diffusion. The same plume behavior would be observed in subsurface or shunted discharges except that, because of its density, the plume would be confined to that part of the water column deeper than the point of discharge. In addition to the main or lower plume, a visible or upper plume is also formed. This results from turbulent mixing of the lower plume with seawater as it descends. The upper plume contains only a small fraction (less than 10 percent of the dishcarged material.

Fractionation of the contaminants may occur during any of the stages in dispersion, depending on whether the materials are soluble or solids heavier or lighter than seawater. Neutrally buoyant solids

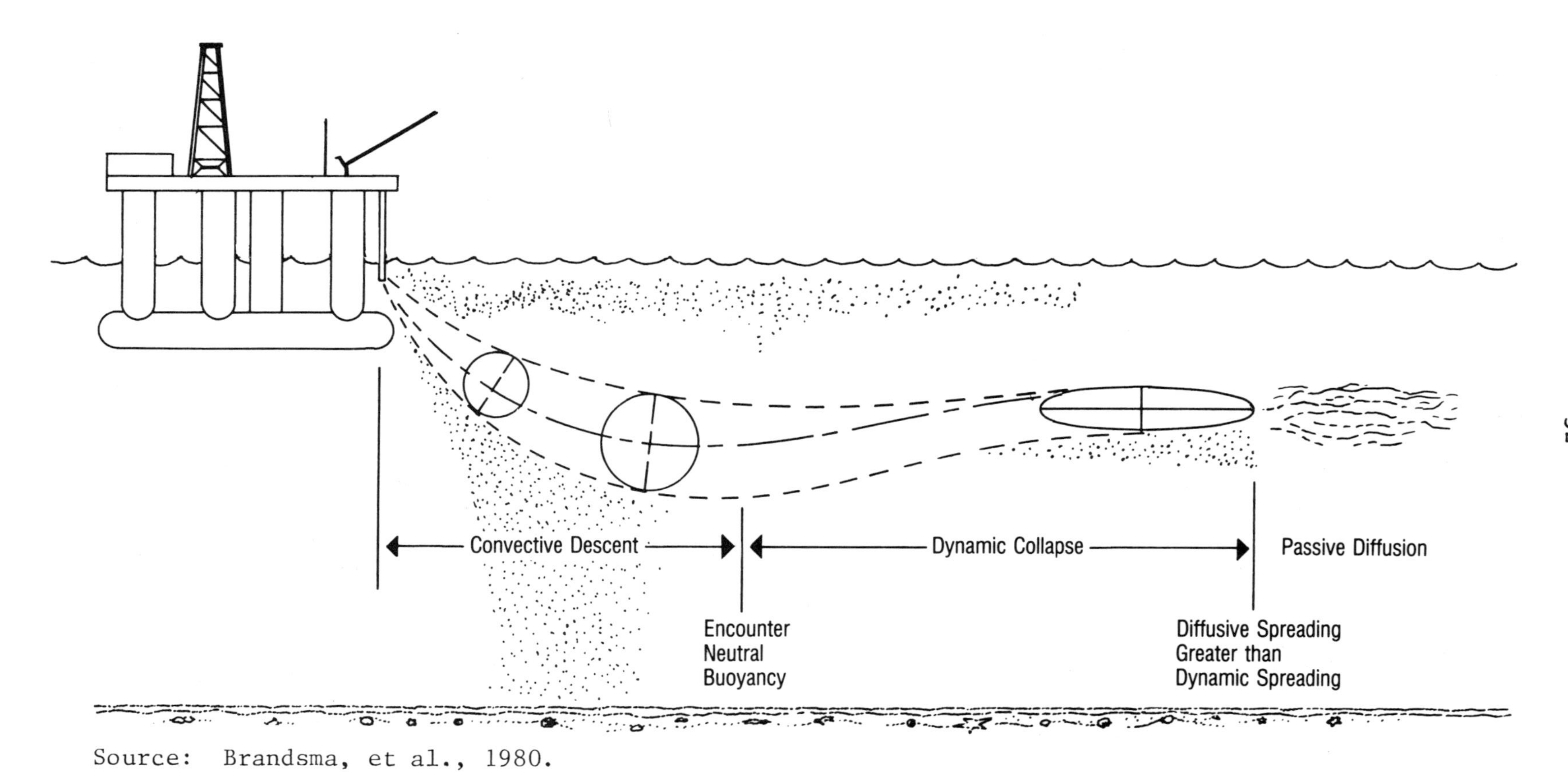


Source: Brandsma, et al., 1980.

FIGURE 2 Idealized Jet Discharge

may settle to an appropriate density discontinuity, where they may be transported at an intermediate depth for long distances from the site of disposal.

Rapid flocculation and aggregation of the clay-sized particles occur when drilling fluids encounter seawater. These finely divided particles are dispersed in drilling fluids by the electrical charges between particles and lignosulfonate, which is used in drilling as a deflocculant. This suspension is destabilized by decreasing lignosulfonate concentration and by particle ion exchange with seawater electrolytes, particularly polyvalent ions. The rate at which agglomeration occurs depends on the frequency of collisions and on the efficiency of particle contacts. Particles in suspension collide with each other as the result of two mechanisms of particle movement. Particles move relative to each other because of thermal energy (brownian motion) and because of the turbulence of the seawater. In the oceans, if colloids are large (0.1-1$\mu$M) or the fluid shear rate is high (10-200 $cm^2/s$), the relative motion of particles created by turbulence far exceeds brownian motion and thus flocculation depends essentially on turbulence.

The decrease in number of particles in a well-mixed (velocity gradient) fluid from agglomeration is expressed by the equation (Stumm and Morgan, 1970):

with $$\mathrm{Ln}\,\frac{n}{N_o} = -K_o t\downarrow, \quad K_o = \alpha_o \frac{n}{\sigma} V_m \frac{du}{dz}$$

where n is the number of particles, $N_o$ is the initial number of particles, $V_m$ is volume of total solid mass suspended per volume of medium, $\alpha_o$ is the fraction of collisions leading to permanent agglomeration, t is time, and du/dz is the velocity gradient. In a practical way this expression indicates that, for a medium containing $10^7$ particles per $cm^3$, of diameter approximating 1$\mu$m, V becomes approximately $5 \times 10^{-6}$ $cm^3/cm^3$. For $\alpha = 1$ and agitation characterized by a velocity gradient of 10/s (equivalent to a "medium" stirred beaker, but generally less than ambient ocean turbulence), $K_o$ is of the order $1.5 \times 10^{-5}$/s. In this example, half the particles would agglomerate in a period of approximately 4.5 h. The high dilution rates in the ocean will reduce the rate of flocculation downstream by reducing the concentration of particles. Materials will be precipitated for extended periods of time because of the complex interactions of microscopic particles suspended in highly ionic solutions.

In the case of drilling fluids, most of the discharged material (barite, flocculated clays, and formation solids) sinks to the bottom near the well site with the distance from the discharge point (within 500 m for most OCS areas) dependent on depth of water, lateral transport, particle size, and density of material (Ayers et al., 1980a; Ayers et al., 1980b; Ray and Meek, 1980; Trefry et al., 1981; Trocine and Trefry, 1982). How long the settled material remains at the well site depends on environmental factors (such as water depth and energy regime) that govern sediment resuspension, transport, and dispersion.

A smaller portion of the discharged material, less than 10 percent of the solids and some of the water and soluble components (Ayers et al., 1980b), remains in the water column. This fraction of the discharge, which breaks away from the fast-descending main plume to form the upper plume, is transported away from the well by ambient currents. The fate of the upper plume (as well as the main plume if it reaches neutral buoyancy before encountering the seabed) depends largely on oceanic dispersion processes. Diffusion as a physical process in the environment has been the subject of considerable study over the past several decades. Batchelor (1952), Ichye (1963), Joseph and Sender (1958), Richardson (1926), and Stommel (1949), developed the theory of this process. In the 1950s the use of fluorescent dyes to measure diffusion directly (Seligman, 1955; Moon et al., 1957; Prichard and Carpenter, 1960) provided many data on the rates of dispersion (dilution) under different oceanographic conditions. In a summary of dye diffusion studies, Okubo (1968) compared the horizontal variance in dye concentration in the upper mixed layer of four geographically separated areas of the continental shelf and two estuaries (Figure 3). These experiments show that, regardless of the detailed oceanographic conditions, variance[4] exhibits a general trend. Specifically, it increases with time by power between 2 and 3 in different scale diffusion fields, current regimes, and sea surface conditions. Thus, local variability appears to have a relatively minor influence on dispersion compared to generic hydrodynamic processes, those common to continental shelf and estuarine waters.

In one of the classical experiments in open ocean diffusion, Folsom and Vine (1957) measured the spread of a radioactive tracer over a horizontal area of 40,000 $km^3$ in 40 days. During this time it mixed vertically through 60 m. This mixing corresponded to eddy diffusivities of $10^7$ $cm^2/s$ in the horizontal and of 1 $cm^2/s$ in the vertical direction. The effect of bottom friction and resulting mixing by tidal and other bottom currents was not seen in these data because of the great water depths in the area observed.

The energy for mixing dilutes any contaminants by mixing them with uncontaminated water. From this it follows that populations of nonmotile or weakly motile organisms like phytoplankton, zooplankton, larvae, and eggs will be exposed to the contaminated plume while other exposed organisms will be carried out of the plume as diffusion of the plume progresses.

The discussion of dispersion thus far has largely focused on the mixed layer and relates primarily to dilution of the buoyant plume. This plume represents less than 10 percent of the discharged material in drilling fluids and cuttings. The bulk of the material, which

---

[4]A statistical term denoting the mean of the squares of variations from the mean of a frequency distribution. In this report, variance refers to the square of the mean of dye concentration variation in a horizontal field from the peak concentration at the center.

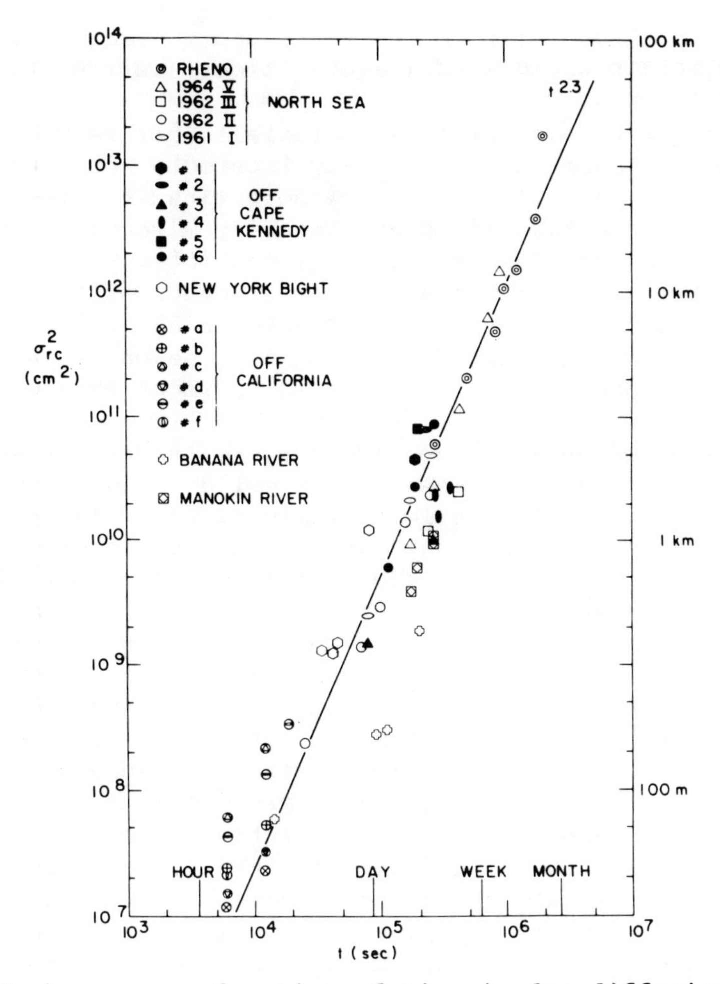


FIGURE 3 Variance as a function of time in dye diffusion experiments.

rapidly sinks to the bottom, is dispersed or remains in place depending on the eddy diffusivity of the bottom boundary layer. A reasonable model for the eddy diffusivity of the bottom boundary layer has been developed (Businger and Arya, 1974; Grant and Glenn, in press; Long, 1982):

$$V_t = \frac{\kappa_o U_* Z}{\phi_m} e^{Z/h}$$

where $\kappa_o$ is von Karmens constant (equals 0.4), $U_*$ is shear velocity, $\phi_m$ is stratification correction, h is boundary layer thickness and Z is roughness length established by measuring near-bottom average velocity profiles.

Thus, as this expression indicates, mixing is affected by all of the following: the depth of discharge (since eddy diffusion has a maximum); how the sediment is distributed over the water column when resuspended; how stratified the flow is (since stratification decreases mixing, thus increasing $\phi_m$); and the boundary shear stress $U_*$.

$U_*$ is a critical parameter for sediment transport because it encompasses the interaction among wind (waves), tides (currents), and sediment.

In considering the fates of those materials that reach the bottom, resuspension and transport are of primary interest. For this reason, parameter $U_*$ becomes very important. $U_*$ determines the mean friction on the large-scale flow field and the eddy viscosity. Eddy viscosity is related to eddy diffusivity by a parameter known as Richardson's number, which represents that fraction of the turbulent energy (eddy diffusivity) generated by the shearing stresses that maintain turbulent mixing against the density gradient. $U_*$ is also related to the transport of the resuspended sediment through the mean flow.

On the continental shelf, mean flow usually is determined by the combination of winds and tides. Against a solid boundary the average velocity profile of mean flow is logarithmic (Bowden, 1962). The eddy diffusivity generated by frictional losses from the interaction of flow with the bottom then provides turbulence for the transport of sediments along the bottom boundary layer.

Instantaneous stress, if great enough, resuspends sediment. This stress is associated with the combined wave and current flow, which is coupled through the nonlinear interaction of steady and oscillatory flows. The time-mean-stress determines the friction of the mean flow. Above the wave boundary layer, time-mean-stress is enhanced by wave-current interaction above that value determined solely by current. This enhancement has been established theoretically by Grant and Madsen (1978, 1979) and Smith (1977), and in the field by Cacchione and Drake (1982), Grant et al. (1982), and in the laboratory by Kemp and Simons (1982) and Bakker and Doorn (1978).

The other important feature of the system is the sediment type. The initiation of motion is clearly related to it, as is seabed roughness, a parameter of considerable importance in flow-solid phase interactions. Bed forms develop under combined flows over sand beds. Under low flow conditions, silty-sandy beds are primarily controlled by bioturbation (Grant et al., 1982), which influences seabed roughness by causing mounds and furrows and adhesion in the sediment. The fates of drilling fluids in different sediments may vary greatly.

## FATES OF DRILLING FLUIDS AND CUTTINGS

The discharge into the ocean of heterogeneous drilling fluids and cuttings results in much fractionation. The biota of the water column are affected by that portion of material that becomes and remains waterborne, the portion that depends on passive diffusion and convection for dispersion. The rates of dispersion are a critical determinant of the fates of these materials in the water column and their effects on the pelagic biota. Effects on the benthos result from that portion of material that settles to the bottom where it can be incorporated into the sediments, resuspended, transported, and dispersed.

## Fates in the Water Column

The preceding brief synopsis of the nature of dispersion in the ocean and the behavior of materials discharged at sea provides a background for considering more specifically the fates of drilling fluids and cuttings that are discharged into the waters of the continental shelf. In the case of cuttings discharges, the relatively large particles settle rapidly near the well. Soluble and particulate fluid additives adhering to the cuttings are to some extent washed off as the larger particles settle. When whole fluid is discharged, most of the material forms a plume which descends rapidly until it encounters the seabed or reaches neutral buoyancy due to water entrainment and solids loss to settling. In addition, a visible or upper plum is formed due to turbulent mixing of the lower plume with seawater. Ayers et al. (1980b) have estimated that the amount of material remaining in the upper plume on discharge is 5 to 7 percent of the total discharge. Under most conditions on the OCS, this portion is of primary concern in considering the fates of materials in the water column. In deeper water (about 80 m or more depending on site conditions), the lower plume will reach neutral buoyancy before encountering the bottom. In this case, both plumes will be of concern in considering water column fates. That portion of the settled material that is resuspended through sedimentary processes will be considered later along with the fates of settled material.

Many field studies have traced the dispersion of the materials contained in the buoyant plume. The studies of Ayers et al. (1980b), Ecomar (1978), Ray and Meek (1980), and Trefry et al. (1981) generally agree with those of Ayers et al. (1980a) and Trocine and Trefry (1982), which will be considered in some detail here as representative studies on the dispersion to be expected under continental shelf conditions. The data obtained in the Ayers et al. (1980a) study are summarized in Table 14, which represents two widely different discharge rates from an exploratory platform at 23 m depth in the Gulf of Mexico. The study was conducted in the summer under calm sea conditions, conditions that did not favor rapid dispersion. For the two discharge rates studied, the rates of change of solids concentration at first decreased rapidly with distance (primarily because of settling), within 45 m in the first study and 152 m in the second; this was followed by a much slower average change in their concentration (primarily because of passive diffusion) of approximately 0.1 mg/l/m or less. Depending upon the discharge rate, the rapid bulk discharge of whole fluids resulted in an initial dispersion of the fluid to between 30 and 50 ppm (solids concentration) within these distances. Further dilution of the plume occurred, approaching ambient total solids concentrations between 350 and 1,500 m from the discharge. At the low discharge rate the transmissivity values persisted below ambient values for somewhat longer than did the values of total solids concentration. This is because fine colloidal material has relatively little mass, but is very effective in scattering light.

The time required for a pollutant to disperse to near-ambient levels is an important parameter in assessing the impact of that pollutant. Since time and distance are related by the velocity of the

TABLE 14 Dilution and dispersion of discharge plumes[a]

| Discharge Rate | Distance from Source (m) | Depth[b] (m) | Suspended Solids[c] (mg/l) | Trans-mittance (%)[c] | Change in Suspended solids (mg/l/m) |
|---|---|---|---|---|---|
| 275 bbl/h[d] | 0 | -- | 1,430,000 | -- | -- |
| | 6 | 8 | 14,800 | -- | -- |
| | 45 | 11 | 34 | 2 | 378.6 |
| | 138 | 9 | 8.5 | 56 | 0.31 |
| | 150 | 9 | 7.0 | 48 | 0.013 |
| | 364 | 9 | 1.2 | 37 | 0.06 |
| | 625 | 9 | 0.9 | 71 | 0.001 |
| | Background | | 0.3-1.9 | 76-85 | |
| 1,000 bbl/h | 0 (Whole mud) | -- | 1,430,000 | -- | -- |
| | 45 | 11 | 855 | 0 | -- |
| | 51 | 12 | 727 | 0 | 21.3 |
| | 152 | 11 | 50.5 | 2 | 6.7 |
| | 375 | 16 | 24.1 | 4 | 0.12 |
| | 498 | 14 | 8.6 | 23 | 0.13 |
| | 777 | 13 | 4.6 | 21 | 0.01 |
| | 878 | 2 | 1.2 | 71 | 0.028 |
| | 957 | 12 | 0.83 | 76 | 0.005 |
| | 1,470 | 11 | 2.2 | 82 | -- |
| | 1,550 | 9 | 1.1 | 82 | -- |
| | Background | | 0.4-1.1 | 80-87 | -- |

[a]Dilution and dispersion of two plumes produced by high-rate high-volume discharges of used chrome lignosulfonate drilling fluid from an offshore exploratory platform in the Gulf of Mexico.
[b]Depth at which highest plume concentration was found.
[c]Maximum solids concentration and minimum transmittance observed at the noted distance.
[d]250 bbl discharged.

SOURCE: Adapted from Ayers *et al*. (1980a).

ambient currents, a plot of the distance from discharge divided by current speed reflects the required time (transport time) to reach given dilution. Ayers et al. (1980a) used this type of analysis to determine the concentration over time of barium in the plume of the rapidly discharged bulk drilling fluids (Figure 4). In this study barium concentration was reduced to between 0.001 and 0.01 percent of its concentration in the drilling fluid within 5 min of discharge. Similar dispersion rates for chromium were also reported.

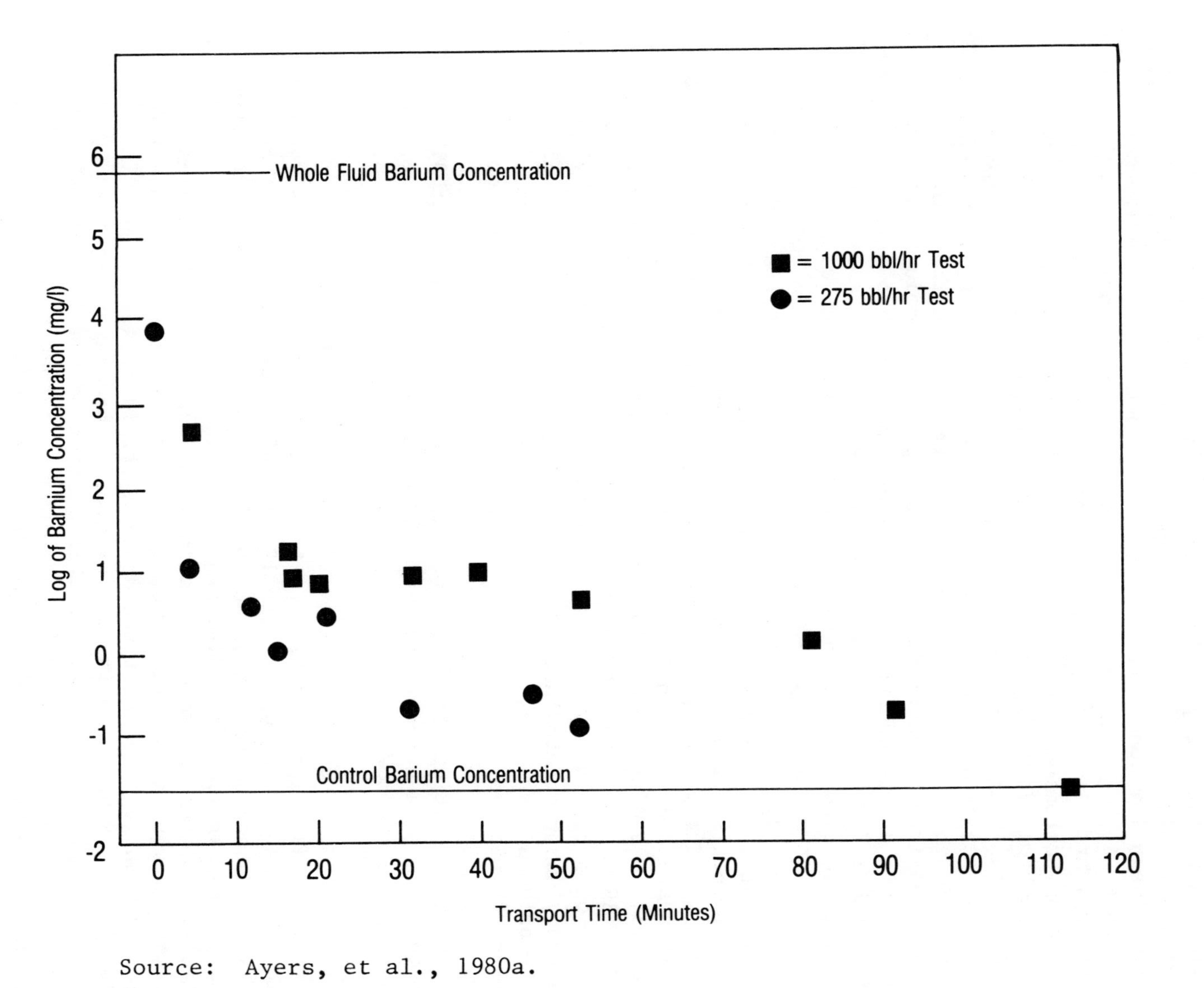


Source: Ayers, et al., 1980a.

FIGURE 4 Barium Concentration as a Function of Transport Time

The studies of Trefry, Trocine, Proni (in press) and Trocine and Trefry (1982) focused on the careful analysis of particulate barium, chromium, and iron in the surface plume as drilling particle tracers. Barium is a particularly useful tracer because it is present in drilling fluids in high concentrations (up to 449,000 μg/g of solids), it has relatively low concentrations in the uncontaminated environment (from 200 to 600 μg/g in near-shore sediments in the northern Gulf of Mexico), and its primary anthropogenic source is drilling fluids. On occasion, chromium has been used as a tracer. Iron is associated with ferrochrome lignosulfonates and other components of drilling fluids and has been found to be a good tracer of bentonite clays with which it associates. Iron was found at levels of 200 ng/l in seawater and 250 μg/g in ambient Gulf of Mexico sediments. Particulate chromium and iron concentrations in the upper water column 200 m downstream of the discharge showed dispersion ratios of $0.5 \times 10^6$ at the surface and $1.5 \times 10^6$ at 10 m depth. These ratios were similar to those for fluid solids and show that chromium and iron closely follow the dispersion of the fluid solids. Particulate barium, on the other hand, showed a dispersion ratio of $1.4 \times 10^6$ at the surface and of $3.0 \times 10^6$ at 10 m, which is two to three times greater than that for the other drilling-fluid solids. Barium thus behaves differently than drilling solids, probably because of its relatively high density. The data indicate a dispersion (dilution) ratio of $10^6$ for drilling solids within a distance of 200 m of a platform with a surface current of 30 to 35 cm/s. Although the fluids were discharged in this case in the form of a fine spray 10 m above the sea surface, the results of this study substantiate those of Ayers et al. (1980a,b). The analytical techniques used here, combined with data comparing metal ratios in drilling fluids and in natural sediments, provide a powerful tool for tracing drilling fluids in the ocean for long distances from the discharge site. Particulate barite was detected in one sample at 3.2 km from the site at a concentration of 750 ng/l in an ambient concentration of 50 to 100 ng/l. The dispersion ratio at this distance was found to be $10^9$.

In the studies described above, components of the drilling fluids investigated had either settled to the bottom or diffused by a factor of $10^6$ to less than 1 mg/l within 100 to 200 m less than 1 h after discharge. A concentration of 30 to 50 mg/l of mud solids was reached within a few minutes after discharge.

At water depths at which the main plume encounters the seabed (50 m or less, depending on site-specific conditions) it may be inferred that less than 10 percent of the drilling fluids (143,000 g/l in the Ayers et al. [1980a] experiments) eventually become transported by the water column plume. Dilution of this portion of the discharge to 30 to 50 mg/l indicates mixing with a volume of seawater 2,500 to 5,000 times greater between 5 and 20 min after discharge. These conditions, which occur soon after discharge, place bounds on the possible extent of exposure of pelagic organisms such as phytoplankton, zooplankton, and micronekton to discharges of drilling fluids.

In deeper water, the lower plume reaches a condition of neutral buoyancy before it encounters the seabed. Thus, both plumes are of interest in water column fate considerations. At this time, virtually all field measurements of drilling fluid solids and soluble component concentrations in the water column have been made on the upper plume. Laboratory measurements and model calculations indicate that concentrations in the lower plume are approximately an order of magnitude higher than in the upper plume (Brandsma amd Sauer, 1983).

## Benthic Fates

Ayers et al. (1980a) showed that over 90 percent of discharged drilling-fluid solids settled directly to the bottom. The distance from the well site and settlement time are primarily a function of current and water depth. Several studies have evaluated the deposition and accumulation of the solids on the seabed (Boothe and Presley, 1983; Bothner, 1982; Dames and Moore, 1978; Ecomar, 1983; EG&G Environmental Consultants, 1982; Gettleson and Laird, 1980; Meek and Ray, 1980; Northern Technical Services, 1983; Trocine et al., 1981). Usually these studies have included analyses of sediment samples for such metals as barium, chromium, iron, lead, and mercury. Of these metals, barium has proven to be the most useful tracer of drilling fluids. Most other metals show moderate to no elevation and are restricted to near-rig (within 125 m) sediments (Boothe and Presley, 1983). In most of the studies total sedimentary concentrations of the elements were determined and no attempt was made to distinguish among metals present as sulfides or hydrous oxides, those sequestered in organic matrices, or those adsorbed on the surfaces of clay particles. Barium would likely persist as particles of $BaSO_4$, although the gradual dissolution of this phase should occur since seawater is undersaturated with respect to $BaSO_4$ (Chow, 1976; Church and Wolgemuth, 1972). In surface waters, biogeochemical scavenging would minimize increases in the concentrations of dissolved barium.

The redistribution and ultimate fates of the settled drilling solids depends upon many environmental factors. The most important factor, as discussed earlier, is the shear velocity, which depends on the shear between the bottom forms and the flow fluid. The sediment type, bioturbation, bottom configuration, and suspended sediment-established stability layers are the primary factors involving the solid phase that influence the shear velocity. The flow field characteristics are determined by the interaction of waves and currents. The highest energy is imparted to the surface water particles by wave action, which induces oscillatory motion. The effect of waves on particle motion decreases with increasing depth, and at depth D it is only a fraction, $\exp(-2\pi^{D/L})$, of that at the surface. L is the distance between a wave crests, i.e., the "wavelength." This factor becomes about 1/2 for a depth of one-ninth of a wavelength, 1/4 for two-ninths of the wavelength and so on. Some motion is contributed to water par- ticles by wave action until D is about half the wavelength. Charac- teristics of typical sea waves are shown in Table 15.

TABLE 15 Typical Sea Waves

| Type of Wave | Period (s) | Wavelength (m) | Velocity (m/s) | Group Velocity (m/s) |
|---|---|---|---|---|
| Ground swell | 15 | 350 | 23.4 | 11.7 |
| Swell | 10 | 156 | 15.6 | 7.8 |
| Ocean waves | 7 | 76 | 10.9 | 5.5 |
| In anchorages | 3 | 14 | 4.7 | 2.3 |

SOURCE: Adapted from Barber and Tucker (1962).

The continental shelf regions are subjected not only to these typical waves, but to storm swells with periods of 14 to 20 s. Such swells impart motion to particles in the water column, and trigger interactions with the bottom sediments over most of the shelf.

The potential effects of storms on sediment transport or movement of drilling solids are very important. Drilling solids may build up for extended periods at certain times of the year, but one major storm event may be sufficient to move the entire layer the solids have formed. Whether this happens depends of course on depth of water, intensity of storms, biological stabilization or destabilization, texture of the ambient sediments and stratification of the flow by suspended sediments. These are the important factors to consider in predicting the fates of deposited drilling solids.

Concern about the sensitivity of hard-substrate epibiota to the physical and toxic effects of drilling fluids has prompted special studies and regulatory restrictions, such as those related to the Flower Garden Banks off the Texas coast. Exploratory drilling activities around the Flower Garden Banks have been monitored to determine the possible effects of these activities on the coral reef ecosystem associated with these banks. In these activities drilling fluids and cuttings had to be shunted to within 10 m of the bottom to protect the banks from the possible plume fallout of materials dispersed in the water column. Studies by Continental Shelf Associates (1975, 1976) and Gettleson (1978) indicate that barium concentrations in ambient sediments vary from 10 to 600 ppm of whole sediment. Postdrilling analysis showed the average barium concentration at 100 m from the drill site to be 3,000 ppm; at 1,000 m from the drill site, the average barium concentration was 1,000 ppm. From 100 to 1,000 m, the concentration decreased inversely with the square root of the distance, and was radially symmetrical around the discharge. The relatively weak currents and low wave energy at the site (bottom currents had a value of about 10 cm/s) resulted in the settlement of most of the solids associated with drilling fluids and cuttings within 1,000 m of the discharge.

Stronger currents were observed in the Tanner Bank area off southern California (Ecomar, 1980; Meek and Ray, 1980). Maximum bottom currents reached 36 cm/s and averaged 21 cm/s. Approximately 863,290 kg of solids were discharged over 85 days. It was inferred from sediment trap data that 12 percent of the solids settled within 50 m. It was estimated that between 44 and 94 percent of this material was in turn transported directly or by resuspension from the drill site by currents, since little accumulation was observed.

Even stronger currents were observed during a study conducted in the Lower Cook Inlet, Alaska (Dames and Moore, 1978). Maximum bottom currents of tidal origin reached 99 cm/s. Sediment trap samples showed that the cuttings and some drilling-fluid particles (barite) were carried initially to the seafloor. However, because of the strong tidal current, dispersion of the settled material was rapid. Television examination of the seafloor at the well site immediately after drilling showed no visible accumulation of cuttings. Barium levels were not elevated. On the other hand, different results were observed in a study conducted in the mid-Atlantic OCS (EG&G Environmental Consultants, 1982). Maximum bottom current reached 18 cm/s and the bottom was too deep (120 m) to be affected by storm waves. In this case, elevated piles of cuttings and sediment barium levels an order of magnitude above ambient levels were both observed in the area of the well site immediately after drilling and 1 year later.

Data from six types of drilling operations, three in shallow water ($\leq$ 34 m) and three in deeper water (76-102 m), showed that water depth was a major controlling factor on bottom deposition (Boothe and Presley, 1983). Detailed sediment analyses revealed that the only component of the drilling discharge remaining in the sediments at a statistically significant level beyond 125 m was barium. A mass balance study showed that only 11.6 percent of the total barium discharged from 25 wells still remained within 500 m of the platform. As can be seen in Table 16, shallow water locations have approximately 10 times less total barium remaining within 500 m. Little (5-10 times) to no elevation of other drilling related metals was seen in near-platform sediments, and only a few stations showed any hydrocarbon elevation.

Two areas of particular interest in oil and gas development are the nearshore Beaufort Sea in the Arctic Ocean and Norton Sound in the Beaufort Sea. These areas differ from the others studied because they are covered with ice for a large part of the year, and drilling in the immediate future will occur in both areas shoreward of the 20-m depth contour. A drilling-fluids discharge study was conducted in the Prudhoe Bay area of the Beaufort Sea in the winter of 1979, in which discharge tests were conducted both below and above the ice (Northern Technical Services, 1981). In the below-ice discharge, rapid dispersion was observed in the 8 m water column, with only loose flocs of

TABLE 16 Mass Balance of Total Excess Sediment Barium Surrounding Offshore Drilling Sites

| Drilling Site[1] | Type[2] | Water Depth (m) | Mode of Discharge[3] | Total Barium Used (TBU) in Drilling Activities ($10^3$ kg) | Mean BAEXAC 0-500m Radius ($g/m^2$)[4] | Total Excess Barium (TEB) in Sediments ($10^3$ kg) within Radius (m)[5] | | | | Percent of TBU within Radius (m) | | | | Ref. |
|---|---|---|---|---|---|---|---|---|---|---|---|---|---|---|
| | | | | | | 500 | 1000 | 2000 | 3000 | 500 | 1000 | 2000 | 3000 | |
| West Cameron 294 | ES | 13 | Surface | 2,414 | 25.8 | 20.3 | -- | -- | 131 | 0.84 | -- | -- | 5.5 | Boothe and Presley, 1983 |
| Vermilion 381 | ED | 102 | Surface | 229 | 28.0 | 22.0 | -- | -- | 193 | 9.6 | -- | -- | 84.0 | " |
| Matagorda 686 | DS | 29 | Surface | 2,334 | 27.5 | 21.6 | -- | -- | 131 | 0.93 | -- | -- | 5.6 | " |
| High Island A-341 | DD | 76 | Surface | 1,518 | 173.0 | 136.0 | -- | -- | 1,093 | 9.0 | -- | -- | 72.0 | " |
| Brazos A-1 | PS | 34 | Surface | 1,041 | 19.2 | 15.1 | -- | -- | 70 | 1.5 | -- | -- | 6.7 | " |
| Vermilion 321 | PD | 79 | Surface | 4,964 | 732.0 | 575.0 | -- | -- | 2,330 | 11.6 | -- | -- | 47.0 | " |

| | | | | | | | | | | | | | | |
|---|---|---|---|---|---|---|---|---|---|---|---|---|---|---|
| High Island A-502 | ES | 55 | Shunted | 127 | 8.8 | 6.9 | 21 | -- | -- | 5.4 | 16.5 | -- | -- | Gettleson and Laira, 1980 |
| Mustang Island A-85 | ED | 75 | Shunted | 820 | 14.0 | 11.0 | 31 | -- | -- | 1.3 | 3.8 | -- | -- | " |
| High Island A-367 | ED | 95 | Surface | 574 | 12.7 | 10.0 | 19 | 46 | 90 | 1.7 | 3.4 | 8.1 | 16.0 | " |
| High Island A-384 | ED | 112 | Shunted | 396 | 143.0 | 117.0 | 129 | 161 | -- | 30.0 | 33.0 | 41.0 | -- | Continental Shelf Associates, 1983 |
| High Island A-389 | ED | 124 | Shunted | 618 | 43.4 | 34.0 | 78 | -- | -- | 5.5 | 12.6 | -- | -- | Gettleson and Laird, 1980 |
| New Jersey (18-3) 684 | ED | 120 | Surface | 443 | 8.2 | 6.4 | 16 | 42 | 86 | 1.5 | 3.7 | 9.5 | 19.0 | EG&G, 1982 |

[1]All drilling sites are in the northwest or north central Gulf of Mexico, except for New Jersey 684, which is located in the western Atlantic 156 km off the coast off New Jersey.

[2]Exploratory (E), development (D), or production (P) in shallow (S) or deep (D) water.

[3]Shunted discharge pipes were located within 10-15 m of the seafloor.

[4]Mean total excess barium in the sediment column areal concentration (BAEXAC) within a 500-meter radius of the drilling site = $TEB_{500}/2\pi(500)^2$.

[5]All TEB data were estimated using the procedure described in Boothe and Presley, 1983 except for the $TEB_{3000}$ values for this study. These $TEB_{3000}$ values were estimated by fitting a power curve of the form $y=ax^b$ to the BAEXAC data for all stations including the 3000m ones. These regressions were significant ($p < 0.01$) for all six drilling sites based on an F test. The range of $R^2$ values was 0.34 to 0.62. The area (representing TEB) under each power curve rotated $360^{\circ}$ and from 500-3000 m radius was integrated. This value was added to the $TEB_{500}$ value to get the $TEB_{3000}$ values given. No actual samples were collected between 500 and 3000 m radii from the drilling sites.. These $TEB_{3000}$ values are included for comparison purposes to give an estimate of the TEB present within 3000 m of these drilling sites.

SOURCE: Boothe and Presley, 1983

drilling materials collecting on the bottom. These were then resuspended and transported by episodic events (probably from changes in barometric pressure) that produced bottom currents of up to 10 cm/s. The determination of trace metals and barium concentrations in the sediment before discharge and 2 to 3 weeks following showed no change. The study of discharge disposal on ice, in which fluids and cuttings were discharged onto the ice surface and allowed to remain until the spring breakup of the ice, caused a broad ultimate dispersal of the materials. A recent open-water study (in an area without ice cover) at the Tern artificial ice island in the Beaufort Sea confirmed the earlier studies (Northern Technical Services, 1983). Drilling fluids and cuttings were prediluted with 30 times their volume of ambient seawater and discharged at the rate of 60 bbl/min into the current impinging on the island. The suspended sediment concentrations were reduced by a factor of 1,000 within 100 m and 15 min of discharge, and no statistically significant patterns of increase in barium, chromium, or lead concentrations were found in the surrounding sediments.

A recent open-water study was made in Norton Sound (Ecomar, 1983) giving further data on the behavior of drilling fluids and cuttings discharged into shallow waters. This study was conducted under extremely adverse weather conditions unlike those of any other similar study, and therefore may serve as a limiting case in analyzing the dispersal of fluids and cuttings in general. Currents at the site were between 18 cm/s at 11 m depth and 80 cm/s at the surface in a southwest direction, and wave-induced motion reached 750 cm/s. These conditions can be compared to those in a Gulf of Mexico study, in which the currents reported were similar, but wave-induced motion was only 9 cm/s. The main difference observed in dispersion patterns was that, in the first study, the wave motion increased the quantity of fluids and cuttings supported by the surface plume longer than at other sites; otherwise, the decrease in ratio of solid concentrations in the fluid to that in the plume was about $5 \times 10^5$ for all cases studied 20 min after the discharge.

Table 17 summarizes the important role of environmental factors in determining the fates of settled materials around a well site. The relatively dispersive energy of the areas is represented by the maximum bottom currents. Water depth is important because it affects how much wave-induced oscillatory currents above the seabed interact with bottom currents to induce sediment resuspension (Grant and Madsen, 1978). In the studies reported in the table, visual evidence of discharged fluids and cuttings was sought through bottom television or submersibles. In Cook Inlet, there was no visual evidence of drilling as soon as the rig left. On Tanner Bank, there was also no visual evidence of discharged material. In the mid-Atlantic, discharges at the well site were still visible 1 year after the rig had moved off location.

Increased barium levels in the sediment immediately after drilling provide another indication of the fates of settled materials. In Cook Inlet, barium levels did not increase. In this area, the energy was so high that the barite particles were rapidly swept away. On Tanner Bank, barium concentrations in the sediment were increased near the

TABLE 17 Effect of Environmental Factors on Study Results

| | Cook Inlet | Tanner Bank | Mid-Atlantic |
|---|---|---|---|
| Environmental factors | | | |
| Maximum bottom current (cm/s) | 99 | 36 | 18 |
| Water depth (m) | 62 | 55 | 120 |
| Study results | | | |
| Visual evidence of discharged material immediately after drilling | No | No | Yes |
| Increased barium levels in sediment immediately after drilling | No | Yes | Yes |

SOURCE: Adapted from Ayers (1981).

well site. In the low-energy mid-Atlantic area, increased barium levels in the sediment around the well site were still found 1 year after drilling. These results show that the length of time the settled material remains concentrated at the well site depends on environmental factors that govern resuspension and dispersion after settlement. This period of time ranges from hours to years.

The processes that govern resuspension of settled particles and their subsequent dispersion are known in theory but have not generally been applied to the fates of deposited drilling fluids and cuttings. Particles are eroded from the seabed when the eddy diffusivity becomes great enough to overcome the adhesive forces of the sediment and the effect of gravity. Eddy diffusivity, as discussed earlier, is a complex function of current velocity, roughness of the sediment, and turbulence induced by waves. The adhesive forces of the sediment are a function of sediment composition, sediment fabric, sedimentation rates, and biological processes, including bioturbation, tube construction, and microbial binding. In some studies, only current velocity and particle size distribution have been used to predict sediment erosion from the seabed (Figure 5). According to these

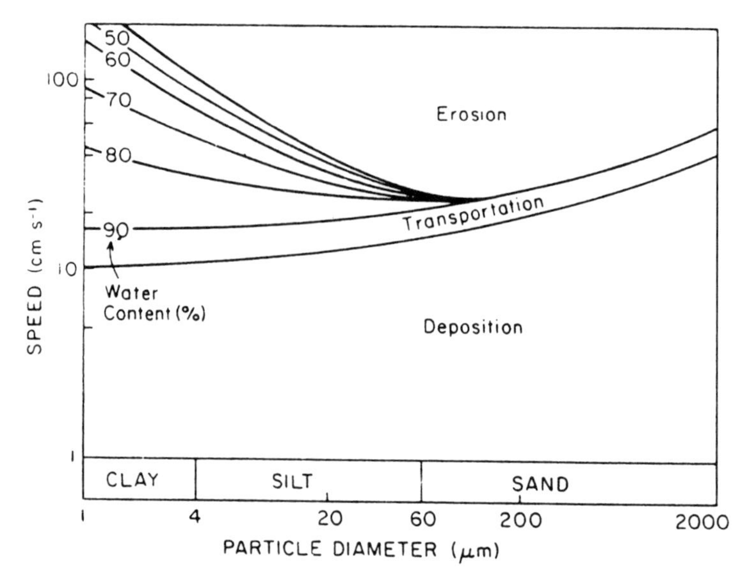


FIGURE 5 Relationship between current speed, particle diameter, and sediment erosion, transport, or deposition (after Kennett, 1982)

criteria, the critical entrainment velocity decreases with decreasing grain size down to fine sand size particles, thereafter varying greatly depending on the cohesiveness and consolidation of sediments. It is now well established that shear stress (closely related to eddy viscosity), and not velocity, is the variable of interest since it takes into account the seabed roughness factor and turbulent motion from waves; velocity alone does not. Figure 5 indicates that a velocity of 100 cm/sc or more is required to erode consolidated clay-sized sediments from the seabed. Most of the time on the shelf, mean flow is insufficient to move sediments, but waves resulting in bottom orbital velocities, which cause sufficient shear stress to erode sediments, are common. Storm waves during seasons of heavy weather are the main determinant of sediment transport in many continental shelf environments.

Both settling and erosion of particles on the seabed is related to a nondimensional fall diameter $S_*$ of a sediment particle in the conventional Shields' diagram (Madsen and Grant, 1976). The term $S_*$ is a function of both particle diameter, particle density, and density of the fluid medium. While the Shields' diagram fails to consider seabed roughness, biological effects, and sediment mixtures, it is useful in showing that particles of considerably different densities (e.g., bentonite, barite, shales, and sandstones), such as those in drilling fluids and cuttings, will undergo selective dispersion under prevailing continental shelf current (10 to 50 cm/s) and wave (5 to 15 s) conditions. Furthermore, the effects of organisms, from microbes to large mammals, may increase or decrease the critical eddy viscosity by changing bottom roughness and sediment cohesiveness (Grant el al., 1982; Nowell et al., 1981).

# REFERENCES

Ayers, R.C., Jr. 1981. Fate and effects of drilling discharges in the marine environment. Proposed North Atlantic OCS Oil and Gas Lease Sale 52. Statement delivered at public hearing, Boston, Mass., November 19, 1981. Bureau of Land Management, U.S. Department of the Interior.

Ayers, R.C., Jr., T.C. Sauer, Jr., D.O. Steubner, and R.P. Meek. 1980a. An environmental study to assess the effect of drilling fluids on water quality parameters during high rate, high volume discharges to the ocean. In: Proceedings of a Symposium on Research on Environmental Fate and Effects of Drilling Fluids and Cuttings. Washington, D.C.: Courtesy Associates. Pp. 351-391.

Ayers, R.C., Jr., T.C. Sauer, Jr., R.P. Meek, and G. Bowers. 1980b. An environmental study to assess the impact of drilling discharges in the mid-Atlantic. I. Quantity and fate of discharges. In: Proceedings of a Symposium on Research on Environmental Fate and Effects of Drilling Fluids and Cuttings. Washington, D.C.: Courtesy Associates. Pp. 382-418.

Barber, N.F., and M.J. Tucker. 1962. Wind waves. In: M.N. Hill (ed.), The Sea. Vol. 1. New York: Interscience. Pp. 664-699.

Batchelor, G.K. 1952. Diffusion in a field of homogeneous turbulence. Proc. Camb. Philos. Soc. 48:345-362.

Bakker, W.T., and T.H. Doorn. 1978. Near bottom velocities in waves with a current. In: Proceedings of a Coastal Engineering Conference. Pp. 1394-1413.

Boothe, P.N., and B.J. Presley. 1983. Distribution and behavior of drilling fluid and cuttings around Gulf of Mexico drill sites. Draft final report. API Project No. 243. American Petroleum Institute, Washington, D.C.

Bothner, M.H., et al. 1982. The Georges Bank monitoring program. Analysis of trace metals in bottom sediments. First year final report to the New York OCS Office, Minerals Management Service, U.S. Department of the Interior. Interagency Agreement No. AA851-IA2/8. Geological Survey, U.S. Department of the Interior, Woods Hole, Mass.

Bowden, K.F. 1962. Turbulence. In: M.N. Hill (ed.), The Sea. New York: Interscience. Pp. 802-825.

Brandsma, M.G., L.R. Davis, R.C. Ayers, Jr., and T.C. Sauer, Jr. 1980. A computer model to predict the short-term fate of drilling discharges in the marine environment. In: Proceedings of a

Symposium on Research on Environmental Fate and Effects of Drilling Fluids and Cuttings. Washington, D.C.: Courtesy Associates. Pp. 588-610.

Brandsma, M.G., and R.C. Sauer. 1983. The OOC model: prediction of short term fate of drilling fluids in the ocean. Part two: model results. In: Proceedings of Minerals Management Service Workshop on Discharges Modeling, Santa Barbara, Calif., February 7-10, 1983.

Businger, J.A., and S.P.S. Arya. 1974. Height of the mixed layer in the stably stratified planetary boundary layer. In: Advances in Geophysics. New York: Academic Press. Pp. 73-92.

Cacchione, D.A., and D.E. Drake. 1982. Measurements of storm generated bottom stresses on the continental shelf. J. Geophys. Res. 87(C3):1952-1961.

Chow, T.J. 1976. Barium in Southern California coastal waters: a potential indicator of marine drilling contamination. Science 193:57-58.

Church, T.M., and K. Wolgemuth. 1972. Earth Planet. Sci. Lett. 15:35-44.

Colonell, J.M. (ed.). 1981. Port Valdez, Alaska: environmental studies 1976-1979. Occasional Publication No. 5. Institute of Marine Science, University of Alaska, Fairbanks, Alaska. 373 pp.

Continental Shelf Associates. 1975. East Flower Garden Bank environmental survey. Rep. No. 1: Pre-drilling environmental assessment, Vol. 1, II. Rep. No. 2: Monitoring program and post-drilling environmental assessment. Vols. I, II, III, IV. Reports submitted to Mobil Oil Corp. for lease OCS-G2759.

Continental Shelf Associates. 1976. Pre-drilling survey report. Results of gravity core sediment sampling and analysis for barium. Post-drilling survey report. Results of gravity core sediment sampling and analysis for barium. Block A-85, Mustang Island Area, East Addition. Reports submitted to Conoco, Inc.

Continental Shelf Associates. 1982. Environmental monitoring program for platform "A," Block A-85, Mustang Island Area, East Addition, near Baker Bank, March 1978 to March 1981. Report submitted to Conoco, Inc.

Dames & Moore, Inc. 1978. Drilling fluid dispersion and biological effects study for the lower Cook Inlet C.O.S.T. well. Report submitted to Atlantic Richfield Co. Dames & Moore, Inc., Anchorage, Alaska. 309 pp.

Ecomar, Inc. 1978. Tanner Bank mud and cuttings study. Conducted for Shell Oil Company, January through March 1977. Ecomar, Inc., Goleta, Calif. 495 pp.

Ecomar, Inc. 1980. Maximum mud discharge study. Conducted for Offshore Operators Committee, Environmental Subcommittee, under direction of Exxon Production Research Co. Ecomar, Inc., Goleta, Calif. 495 pp.

Ecomar, Inc. 1983. Mud dispersion study, Norton Sound Cost Well No. 2. Conducted for Arco Alaska, Inc., Anchorage, Alaska. 91 pp.

EG&G Environmental Consultants. 1982. A study of environmental effects of exploratory drilling on the Mid-Atlantic Outer Continental Shelf--final report of the Block 684 Monitoring Program. EG&G Environmental Consultants, Waltham, Mass. Available from Offshore Operators Committee, Environmental Subcommittee, P.O. Box 50751, New Orleans, LA 70150.

Folsom, T.R., and A.C. Vine. 1957. On the tagging of water masses for the study of physical processes in the oceans. In: Natl. Acad. Sci.-Natl. Res. Counc. Publ. 551:121-132.

Gettleson, D.A. 1978. Ecological impact of exploratory drilling: a case study. In: Energy/Environment 1978. Society of Petroleum Industry Biologists Symposium, 22-24 August, 1978, Los Angeles, Calif. 23 pp.

Gettleson, D.A., and C.E. Laird. 1980. Benthic barium in the vicinity of six drill sites in the Gulf of Mexico. In: Proceedings of a Symposium on Research on Environmental Fate and Effects of Drilling fluids and Cuttings. Washington, D.C.: Courtesy Associates. Pp. 739-788.

Grant, W.D., and S.M. Glenn. In press. Continental shelf bottom boundary layer model: theoretical model. Vol. I. Woods Hole Oceanographic Institution Technical Memorandum. 160 pp.

Grant, W.D., S.M. Glenn, and A.J. Williams. In press. Near bottom velocity profile measurements under combined wave and current flows. Comparison with theory. Woods Hole Oceanographic Institution Technical Memorandum. 160 pp.

Grant, W.D., L.F. Boyer, and L.P. Sanford. 1982. The effects of bioturbation on the initiation of motion of intertidal sands. J. Mar. Res. 40:659-677.

Grant, W.D., and S.M. Glenn. 1983. Continental shelf bottom boundary layer model. Vol. I: Theoretical model development. Final report to the American Gas Association PR-153-125. Department of Ocean Engineering, Woods Hole Oceanographic Institution. 163 pp.

Grant, W.D., and O.S. Madsen. 1979a. Combined wave and current interaction with a rough bottom. J. Geophys. Res. 84:1797-1808.

Grant, W.D., and O.S. Madsen. 1979b. Bottom friction under waves in the presence of a weak current. NOAA TM ERL/MESA-29. 150 pp.

Grant, W.D., and O.S. Madsen. 1982. Moveable bed roughness in unsteady oscillatory flow. J. Geophys. Res. 87(C1):469-481.

Gross, M. Grant. 1975. Trends in waste solid disposal in U.S. coastal waters, 1968-1974. In: Thomas M. Church (ed.). Marine Chemistry of the Coastal Environment. American Chemical Society ACS Symposium Series 18. 710 pp.

Hill, W.N. (ed.). 1962. Physical oceanography. In: The Sea. Vol. 1. New York: Interscience. 864 pp.

Hood, D.W., B. Stevenson, and L.M. Jeffery. 1958. Deep sea disposal of industrial wastes. Ind. Eng. Chem. 50:885-888.

Ichye, T. 1963. Oceanic turbulence. Technical Paper No. 2. Prepared for Office of Naval Research by Florida State University. 200 pp.

Joseph, J., and H. Sendner. 1958. Uber die horizontale diffusion im meere. Dtsch. Hydrogr. Z. 11:49-77.

Judson, S., and D.F. Ritter. 1964. Rates of regional denudation in the United States. J. Geophys. Res. 64:3395-3401.

Kemp, P.H., and R.R. Simons. 1982. The interaction between waves and a turbulent current: waves propogating with the currents. J. Fluid Mech. 116:227-250.

Kennett, J.P. 1982. Marine Geology. Englewood Cliffs, N.J.: Prentice Hall. 813 pp.

Ketchum, B.H., and W.L. Ford. 1952. Rate of Dispersion in the Wake of a Barge at Sea. Trans. Am. Geophys. Union 33(5):680-684.

Long, C.E. 1981. A simple model for time-dependent stably stratified turbulent boundary layers. Special Report No. 95. Department of Oceanography, University of Washington, Seattle. 170 pp.

Madsen, O.S., and W.D. Grant. 1976. Sediment transport in the coastal environment. Rep. No. 209. Ralph M. Parsons Lab., Massachusetts Institute of Technology. 105 pp.

Meek, R.P., and J.P. Ray. 1980. Induced sedimentation, accumulation, and transport resulting from exploratory drilling discharge of drilling fluids and cuttings. In: Proceedings of a Symposium on

Research on Environmental Fate and Effects of Drilling Fluids and Cuttings. Washington, D.C.: Courtesy Associates. Pp. 259-284.

Moon, F.W., Jr., C.L. Bretschneider, and D.W. Hood. 1957. A method for measuring eddy diffusion in coastal embayments. Inst. Mar. Sci. 1(Pub. Univ. Tex.)IV:14-21.

Northern Technical Services. 1981. Beaufort Sea drilling effluent disposal study. Prepared for the Reindeer Island stratigraphic test well participants under the direction of SOHIO Alaska Petroleum Company.

Northern Technical Services. 1983. Open water drilling effluent disposal study, Tern Island, Beaufort Sea, Alaska. Conducted for Shell Oil Co. 87 pp.

Nowell, A.R.M., P.A. Jumars, and J.E. Ekman. 1981. Effects of biological activity on the entrainment of marine sediments. Mar. Geol. 42:133-153.

Okubo, A. 1968. A new set of oceanic diffusion diagrams. Chesapeake Bay Inst. Tech. Rep. 38.

Okubo, A. 1971. Horizontal and vertical mixing in the sea. In: D.W. Hood (ed.), Impingement of Man on the Ocean. New York: Wiley-Interscience. 738 pp. Pp. 89-168.

Petrazzuolo, G. 1981. Preliminary report. An environmental assessment of drilling fluids and cuttings released onto the outer continental shelf. Vol. 1: Technical assessments. Vol. 2: Tables, figures and Appendix A. Prepared for Industrial Permits Branch, Office of Water Enforcement and Ocean Programs Branch, Office of Water and Waste Management, U.S. Environmental Protection Agency, Washington, D.C.

Prichard, D.W., and J.H. Carpenter. 1960. Measurements of turbulent diffusion in estuarine and inshore waters. Bull. Int. Assoc. Sci. Hydrol. 20:37-50.

Ray, J.P. 1979. Offshore discharge of drilling muds and cuttings. In: Outer Continental Shelf Frontier Technology: Proceedings of a Symposium. Washington, D.C.: National Academy of Sciences.

Ray, J.P., and R.P. Meek. 1980. Water column characterization of drilling fluids dispersion from an offshore exploratory well on Tanner Bank. In: Proceedings of a Symposium on Research on Environmental Fate and Effects of Drilling Fluids and Cuttings. Washington, D.C: Courtesy Associates. Pp. 223-258.

Ray, J.P., and E.A. Shinn. 1975. Environmental effects of drilling muds and cuttings. In: Environmental Aspects of Chemical Use in Well Drilling Operations. EPA-560/1-75-004. Washington, D.C.: U.S. Environmental Protection Agency. Pp. 533-545.

Richardson, L.F. 1926. Atmospheric diffusion shown on a distance-neighbor graph. Proc. R. Soc., Lond. A110:709-727.

Seligman, H. 1955. International Conference on the Peaceful Uses of Atomic Energy. Paper No. 419.

Smith, J.D. 1977. Modeling of sediment transport on continental shelves. In: The Sea. Vol. 6. New York: Wiley-Interscience.

Stommel, H. 1949. Horizontal diffusion due to oceanic turbulence. J. Mar. Res. 8:199-225.

Stumm, W., and J.J. Morgan. 1970. Aquatic Chemistry. New York: Wiley-Interscience. 583 pp.

Sverdrup, H.U., M.W. Johnson, and R.H. Fleming. 1942. The Oceans. Englewood Cliffs, N.J.: Prentice Hall. 1060 pp.

Trefry, J.H., R.P. Trocine, and D.B. Meyer. 1981. Tracing the fate of petroleum drilling fluids in the northwest Gulf of Mexico. In: Conference Record: Oceans 81, Boston, Mass. Vol. 2. Available from Marine Technology Society, Washington, D.C. Pp. 732-736.

Trefry, J.H., R.P. Trocine, and J.R. Proni. In press. Drilling fluid discharges into the marine environment. Presented at Third International Ocean Disposal Symposium, Woods Hole Oceanographic Institution, October 12-16, 1981. In: Wastes in the Ocean. New York: John Wiley & Sons.

Trocine, R.P., and J.H. Trefry. 1983. Particulate metal tracers of petroleum drilling fluid dispersion in the marine environment. Environ. Sci. Technol. 17(9).

Yudelson, J.M. 1967. A survey of ocean diffusion studies and data. W.K. Kech Laboratory of Hydraulics and Water Resources Tech. Mem. No. 67-2. California Institute of Technology, Pasadena, Calif.

Zingula, R.P. 1975. Effects of drilling operations in the marine environment. In: Environmental Aspects of Chemical Use in Well Drilling Operations. EPA-560/1-75-004. Washington, D.C.: U.S. Environmental Protection Agency. Pp. 433-448.

# 4
# The Biological Effects of Drilling Discharges

## INTRODUCTION

There are two major environmental concerns about discharging used drilling fluids to the oceans: (1) that these fluids may kill marine organisms, produce harmful sublethal responses in them, or alter ecosystems; and (2) that some of these fluids may contain metals and organic compounds that accumulate in marine organisms to concentrations that could harm them or their consumers, including humans. A substantial body of scientific research addresses these concerns. The purpose of this chapter is to summarize and critically evaluate this literature.

Evaluating the effects of a complex mixture on the marine environment requires many kinds of information. The acute lethal and chronic toxicities of the complex mixture and of its ingredients must be known. Biological responses of marine organisms to sublethal concentrations of the mixture must be not only measured but also evaluated in terms of their ecological significance and implications for human health. Chemical compositions of mixtures resulting in acute and sublethal effects need to be determined. Laboratory studies of the mixture's acute, chronic, and sublethal effects should be interpreted in the context of expected or measured concentrations and exposure durations in the field. Finally, the long-term responses of marine organisms, communities, and ecosystems exposed to these mixtures should be documented in the field. The information available on drilling fluids with regard to all these points provides the basis for evaluating these fluids' effects on the marine environment.

## THE TOXICITIES OF DRILLING FLUID COMPONENTS

A common practice in evaluating the toxicity of such a complex mixture as a drilling fluid is to determine the toxicities of its components in bioassays. The assumption is made that the toxicities of the individual components are approximately additive and that no physical or chemical interactions among ingredients affect the toxicity of the mixture during its formulation or use. These assumptions are probably invalid with regard to used treated drilling fluids. Sprague and Logan (1979) showed that the calculated sum of toxicities of ingredients in a used drilling fluid was not always a good predictor of the acute toxicity of the whole used fluid to freshwater fish. In spite of these limitations, bioassays with drilling-fluid ingredients are likely to be useful in identifying the relative toxicities of components. If these toxic or physically damaging ingredients are identified, they can in some cases be replaced by less harmful substitutes. The ocean discharge of fluids containing such ingredients also can be regulated.

## Acute Lethal Toxicity

Acute lethal bioassays (usually run over 96 hours) are used to compare the relative acute toxicities of different drilling fluids and drilling-fluid ingredients and the relative sensitivities of different species. They establish the basis for determining quantitative relations between exposures and effects. They are quick and inexpensive and therefore a very popular means of initially screening and ranking the potential hazards of chemicals that might be released to the environment in substantial quantities. They also help determine the ranges of concentrations to be used in studies of chronic and sublethal effects.

Acute lethal bioassays cannot be used alone, however, to predict the environmental effects of discharging drilling fluids to the ocean (see extended discussion of limitations in Chapter 5). In such bioassays, animals often are exposed to drilling-fluid ingredients, drilling fluids, or drilling-fluid fractions in concentrations substantially higher and for much longer than in the field. High concentrations and long exposure times are often needed to produce statistically significant results with a reasonably small number of test animals. If chronic as well as acute bioassay data are generated, it may be possible to extrapolate (using application or safety factors) to environmentally more realistic exposure concentrations.

Table 18 presents some of the data available on the acute lethal toxicities of drilling-fluid ingredients to marine and estuarine organisms.

### Major Ingredients

Of the five ingredients that make up more than 90 percent of most water-based drilling fluids, namely, barite, bentonite, lignite, chrome lignosulfonate, and sodium hydroxide (Perricone, 1980), only chrome and ferrochrome lignosulfonates and sodium hydroxide are moderately toxic (LC50 of 100 to 1,000 ppm) to any but the most sensitive species and life stages of marine organisms (Table 18). The 96-h LC50 of NaOH for rainbow trout in fresh water is 105 to 110 ppm (Logan et al., 1973; Sprague and Logan, 1979). The toxic effects of this material are attributed to elevation in pH. Chaffee and Spies (1982) report that adding sufficient NaOH to seawater to increase pH from 7.8 (control) to 8.5 or 9.0 reduced the growth rate and increased the incidence of developmental anomalies in embryos of the starfish *Patiria miniata*. Because of the higher buffer capacity of seawater $2 \times 10^{-3}$ eq/ℓ/pH unit compared to that of fresh water, no significant change in pH

TABLE 18 Acute Toxicity of Drilling Fluid Components to Estuarine and Marine Organisms[a]

| Compound | Bioassay Organism | 96-1 LC50(ppm) | Reference |
|---|---|---|---|
| Aquagel® (Wyoming bentonite) | Oyster *Crassostrea virginica* | >7,500 | Daugherty, 1951 |
| | Shrimp *Pandalus hypsinotus* | 100,000 | Dames and Moore, 1978 |
| | Copepod *Acartia tonsa* | 22,000 | EG&G Bionomics, 1976a |
| | Alga *Skeletonema costatum* | 9,600 | EG&G Bionomics, 1976a |
| Barite (barium sulfate) | Several fish and invertebrates | >7,500 | Daugherty, 1951 |
| | Sailfin molly *Mollieniasis latipinna* | >100,000 | Grantham and Sloan, 1975 |
| | Shrimp *Pandalus hypsonotus* | >100,000 | Dames and Moore, 1978 |
| | Copepod *Acartia tonsa* | 590 | EG&G Bionomics, 1976a |
| | Alga *Skeletonema costatum* | 385-1650 | EG&G Bionomics, 1976a |
| Calcite (calcium carbonate) | Sailfin molly *M. latipinna* | >100,000 | Grantham and Sloan, 1975 |
| Siderite (iron carbonate) | Sailfin molly *M. latipinna* | >100,000 | Grantham and Sloan, 1975 |
| Carbonox® (lignitic material) | Several fish and invertebrates | >7,500 | Daugherty, 1951 |
| Lignite | Sailfin molly *M. latipinna* | >15,000 | Hollingsworth and Lockhart, 1975 |
| Chrome lignosulfonate | Sailfin molly *M. latipinna* | 12,200 | Hollingsworth and Lockhart, 1975 |
| Chrome-treated lignosulfonate | White shrimp *Penacus setiferus* | 465 | Chesser and McKenzie, 1975 |
| Ferrochrome lignosulfonate | Dungeness crab *Cancer magister* | 210[b] | Carls and Rice, 1980 |
| | Dock shrimp *Pandalus danae* | 120[b] | Carls and Rice, 1980 |
| Iron lignosulfonate | White shrimp *Penaeus setiferus* | 2,100 | Chesser and McKenzie, 1975 |
| Cellulosic calcium caronate workover additive | White shrimp *Penaeus setiferus* | 1,925 | Chesser and McKenzie, 1975 |

TABLE 18 (continued)

| Compound | Bioassay Organism | 96-1 LC50(ppm) | Reference |
|---|---|---|---|
| Jelflake® (shredded cellophane) | Several fish and invertebrates | >7,500 | Daugherty, 1951 |
| Impermex® (pregelatinized starch) | Several fish and invertebrates | 500-7,500 | Daugherty, 1951 |
| | Oyster Crassostrea virginica | 3,000 | Daugherty, 1951 |
| Fibertex® (shredded cane fiber) | Several fish and invertebrates | >7,500 | Daugherty, 1951 |
| Mica | Several fish and invertebrates | >7,500 | Daugherty, 1951 |
| Low-molecular-weight polyacrylate | White shrimp Penaeus setiferus | 3,500 | Chesser and McKenzie, 1975 |
| Quebraco (tannin) | Sailfin molly M. latipinna | 158 | Hollingsworth and Lockhart, 1975 |
| Modified hemlock bark extract (tannin) | White shrimp Penaeus setiferus | 265 | Chesser and McKenzie, 1975 |
| Sodium acid pyrophosphate ($Ha_2H_2P_2O_7$) | Sailfin molly M. latipinna | 7,100 | Grantham and Sloan 1975 |
| Oilfos® (Na tetraphosphate) | Several fish and invertebrates | >7,500 | Daugherty, 1951 |
| Quadrafos® (Na polyphosphate) | Several fish and invertebrates | 500-7,500 | Daugherty, 1951 |
| Oil well cement | Several fish and invertebrates | 70-450 | Daugherty, 1951 |
| White lime | Several fish and invertebrates | 70-450 | Daugherty, 1951 |
| Formaldehyde | Pompano Trachinotus carolinus | 25-31 | Birdsong and Avault, 1971 |
| Dowacide G® (79% Na pentachlorophenate) | Sheepshead minnow Cyprinodon variegatus (2 wk. fry) | 0.52 | Borthwick and Schimmel, 1978 |
| | Pinfish Lagodon rhomboides (48-h prolarvae) | 0.066 | Borthwick and Schimmel, 1978 |
| Diesel fuel | Many fish and invertebrates | 0.1-1,000 | Neff and Anderson, 1981 |

[a]IMCO et al. (1969) defines LC50 toxicities as follows: very toxic, <1ppm; toxic, 1-100 ppm; moderately toxic, 100-1,000 ppm; slightly toxic 1,000-10,000 ppm; practically nontoxic, >10,0900 ppm.
[b]144-h LC50.

occurs when drilling fluid is discharged to the ocean. Based on the evidence to date, barite, bentonite, and lignite can be classified as practically nontoxic (having LC50 values greater than 10,000 ppm). Although EG&G Bionomics (1976a,b) reported that barite was moderately toxic (96-h LC50 of 385 to 1,650 ppm) to a copepod, *Acartia tonsa*, and an alga, *Skeletonema costatum*, mortality in these bioassays can probably be attributed to physical abrasion by suspended barite particles and not to chemical toxicity.

## Chromium

Drilling fluids may contain chromium in a variety of chemical forms, but mostly complexed with lignosulfonate. It is generally believed that virtually all the chromium in a drilling fluid that has been used for an extended period will be in the trivalent state, even though it may have been added as inorganic hexavalent chromium (Skelly and Dieball, 1969). This may not always be the case. In a chromate-treated chrome lignosulfonate fluid maintained at a pH of 9 to 11, hexavalent chromium may persist for a long time at room temperature (23 to 24°C). At higher temperatures typical of those encountered near the drill bit (50 to 120°C), chromate is reduced rapidly to trivalent chromium and becomes complexed with the lignosulfonate molecule (Skelly and Dieball, 1969). Chromium-lignosulfonate complexes are quite stable at normal operating temperatures and pHs, and chromium is not readily released (McAtee and Smith, 1969). At temperatures above about 150°C, chrome lignosulfonates lose their ability to thin drilling fluids, primarily because of thermal degradation of the organic portion of the molecule and not because of a change in the physical or chemical form of the chromium (Carney and Harris, 1975).

Thermodynamic calculations indicate that when chromate-treated chrome lignosulfonate drilling fluid is discharged to the ocean and diluted with seawater any chromium ions in the trivalent state, Cr(III), should be transformed to the hexavalent state Cr(VI), and any Cr(VI) discharged with the fluid should remain in that valence state (Cranston and Murray, 1980; Nakayama et al., 1981a,b,c; van der Weijden and Reith, 1982). The oxidation of Cr(III) to Cr(VI) occurs very slowly in normally oxygenated seawater, however, and Cr(III) tends rapidly to adsorb to or complex with suspended organic material and clay. In the complexed state, Cr(III) is very resistant to oxidation. Manganese oxides in seawater or sediments may accelerate the oxidation of Cr(III) to Cr(VI), while oxidizable materials like $H_2S$ and many natural organic compounds readily reduce Cr(VI) to Cr(III) (Nakayama et al., 1981b; Smillie et al., 1981). Cr(VI) is also reduced rapidly to Cr(III) in anoxic sediments (van der Weijden and Reith, 1982). As a result of these interactions, seawater may contain dissolved chromium in the form of Cr(III), Cr(VI), and organically bound chromium (Nakayama et al., 1981c).

Trivalent chromium salts are not very soluble in seawater and have low toxicities (see, for example, Oshida et al., 1981). Most species of marine animals are much more sensitive to hexavalent chromium salts than to trivalent salts, although species vary greatly in their sensitivities to the first (Eisler and Hennekey, 1977; Frank and Robertson, 1979; Reish et al., 1976). Some species appear to be equally sensitive to Cr(VI) and Cr(III). The 48-h LC50s of Cr(VI), as $Na_2CrO_4$, and of Cr(III), as $CrO_3$, are 16.37 and 19.27 ppm respectively for the marine copepod Acartia clausi (Moraitou-Apostolopoulou and Verriopoulos, 1982). The reported LC50 value for $CrO_3$ is about 385 times higher than the value at which this Cr(III) salt is soluble in seawater, so the bioassay has no environmental meaning. Published values for the acute lethal toxicity (usually 96-h LC50) of inorganic hexavalent chromium salts to marine animals typically fall in the range of 0.5 to 250 mg/l. Polychaete worms are very sensitive; teleost fish are not.

Slightly soluble hexavalent chromium salts, such as calcium chromate and zinc chromate, are carcinogenic in mammals following tracheal inhalation and intramuscular or intrapleural injection (Norseth, 1981). Chromates also show evidence of genetic toxicity in several in vitro tests. Trivalent chromium salts have shown little or no evidence of carcinogenicity and genetic toxicity. When ingested in small amounts by marine animals or man chromates would be reduced to the trivalent state by organic materials in the digestive tract, and therefore would not represent an important carcinogenic hazard.

In used chrome lignosulfonate drilling fluids, the proportion of total or dissolved chromium present in ionized inorganic form is not known (Liss et al., 1980). Since most of the chromium is associated with the lignosulfonate, or clay fractions of the fluid or both, data on the toxicity of ionic chromium species do not accurately indicate the contribution of chromium to the toxicity of drilling fluids.

## Hydrocarbons

Diesel fuel (No. 2 fuel oil) sometimes is added to water-based drilling fluids to improve the lubricating properties of the fluid when drilling a slanted hole. As much as 2 to 4 percent diesel fuel may be added under some circumstances. Because much of the added diesel fuel quickly becomes adsorbed to particles in the drilling fluids, discharging these fluids to the ocean rarely results in an oil sheen on the surface. In some circumstances, a "pill" of diesel fuel or oil-based fluid is used to help free stuck pipe. This pill may or may not be kept separate from the bulk fluid system, recovered, and disposed of onshore. Even when the pill is kept separate, a small amount of diesel fuel from the pill may become mixed with the bulk drilling fluid.

There is growing evidence that diesel fuel may contribute significantly to the toxicity of drilling fluids that contain it. Conklin et al. (in press) reported a statistically significant inverse relationship ($r = -0.58$, $p < 0.05$) between the 96-h LC50 of 18 drilling fluid samples collected at different depths from an exploratory well in Mobile Bay, Alabama, to molting grass shrimp Palaemonetes pugio, and the concentrations in these fluids of petroleum hydrocarbons identified as derived from a no. 2 fuel oil. The drilling fluids contained 170 to

8,040 μl of petroleum per liter of whole fluid, and had acute toxicities of 14,560 to 360 μl/l, respectively.

The toxicities of crude and refined petroleums to marine organisms have been studied extensively (see the reviews of Baker, 1976; Malins, 1977; Neff and Anderson, 1981; Rice et al., 1977). The acute toxicities of different petroleums to different species of marine organisms are extremely variable. Most 96-h LC50s fall in the range of 1 to 1,000 mg of oil per liter. Some very sensitive larval and early life stages of marine animals may have LC50s of about 0.1 to 1 mg/l. Sublethal responses to petroleum hydrocarbons, especially behavioral modifications, have been reported following acute exposure to petroleum concentrations in the low microgram per liter (parts per billion) range. Because the most toxic major components of petroleum are the light aromatics (i.e., benzenes, naphthalenes, and phenanthrenes) and closely related heterocyclic compounds, the acute toxicity of a particular crude, refined, or residual petroleum product is usually directly correlated with the concentration in it of these compounds.

No. 2 diesel fuel is a petroleum hydrocarbon distillate of moderate volatility used in medium and high speed engines in industrial and heavy mobile service (such engines commonly power drilling rigs). No. 2 diesel fuel has a boiling range of 350-700$^{o}$F. While the properties of U.S. diesel fuels are generally defined by ASTM Specification D-975, the exact composition of a sample of No. 2 diesel fuel will vary with the source, refining techniques, seasonal climatic requirements and the demand for other products at the processing facility. No. 2 diesel fuel may include 20-40 percent cracked components and 15 to 50 percent by weight aromatic hydrocarbons. No. 2 diesel fuel may also contain a variety of additives, used in providing properties such as oxidation inhibition, dispersancy, corrosion protection, cetane improvement, and anti-static protection. The toxicity and environmental effects of additives to diesel fuel are evaluated, as an element of registering the products with the EPA.

Most of the aromatic hydrocarbons in diesel fuels are benzenes and naphthalenes, with much smaller amounts of polycyclic aromatic hydrocarbons (aromatics containing three or more fused aromatic rings.) An American Petroleum Institute reference No. 2 diesel fuel for biological effects research contained 38.2 percent by weight aromatic hydrocarbons (Neff and Anderson, 1981). A total of 39 percent of the chemical components in the fuel were identified by gas chromatography/mass spectrometry. Of the identified components, 22,000 ppm were benzenes, 65,190 ppm were naphthalenes and biphenyls, and 11,962 ppm were polycyclic aromatic hydrocarbons. A comprehensive review of the fates and effects of petroleum hydrocarbons was published by the National Academy of Sciences in 1975 (National Academy of Sciences, 1975) and is currently being updated (publication is anticipated late in 1983).

A sample of drilling fluid which had been treated with diesel fuel and IMCO Free Pipe from an exploratory well on Georges Bank contained 481 ppm total hydrocarbons, 31.6 percent of which were in the aromatic fraction. Of the identified components 8.7 ppm were benzenes, 20.1 ppm were naphthalenes and biphenyls, 3.1 ppm were polycyclic aromatic hydrocarbons, and 0.8 ppm were dibenzothiophenes (2 ring sulfur heterocyclics). Thus, the hydrocarbon composition of the drilling fluid sample resembled that of diesel fuel except that it was deficient in benzenes which are quite volatile and probably had evaporated during usage or during extraction/clean-up for analysis.

### Biocides

Halogenated phenol biocides, such as Dowicide B and Dowicide G, which have been used as drilling-fluid additives in the past, are quite toxic to benthic invertebrates (Land, 1974; Zitko, 1975). Their use is now prohibited in drilling fluids and all other drilling or production operations on the OCS (Federal Register, 1979). The use of paraformaldehyde is permitted. Its acute and chronic toxicities to marine animals are much lower than those of chlorinated phenol biocides (Table 18). Paraformaldehyde is used in amounts up to 300 g/bbl (about 1,500 ppm). Paraformaldehyde depolymerizes to formaldehyde, the active biocide, upon contact with water. Formaldehyde is suspected of being a carcinogen when administered to rats via inhalation, but the carcinogenicity of traces of formaldehyde in solution to marine organisms is unknown.

### Surfactants

Surfactants are used in small amounts in some drilling fluids to aid the dispersion of poorly soluble components, such as aluminum stearate and gilsonite, in the aqueous phase of the fluid. Polyethoxylated alkyl phenols like Aktaflo-E or Aktaflo-S may be added to drilling fluids at concentrations of 1 to 10 lb/bbl (2,850 to 28,500 ppm) (American Petroleum Institute, 1978). Structurally related polyoxyethylene esters and ethers have acute toxicities in the range of 1 to 40 ppm for Atlantic salmon *Salmo salar* and of 2.5 to 14,000 ppm for the amphipod *Gammarus oceanicus* (Widlish, 1972). Anionic surfactants of the linear alkylate sulfonate and alkyl aryl sulfonate types are sometimes used in drilling fluids. They have acute toxicities (96-h LC50) to freshwater and marine invertebrates and fish of 0.4 to 40 ppm (Abel, 1974). Toxicity increases with decrease in water hardness (or salinity) and decrease in alkyl side chain length.

## Chronic and Sublethal Effects

Relatively little research has been performed on the chronic and sublethal effects of individual drilling-fluid ingredients on marine animals. The research that has been done has focused on barite, hydrocarbons, and various biocides.

### Barite

Several experiments studied the effects of barite on recruiting benthic invertebrates from the plankton to sandy sediments in experimental aquaria receiving unfiltered estuarine water (Cantelmo et al., 1979; Tagatz et al., 1980; Tagatz and Tobia, 1978). The abundance of most species of meiofaunal animals was significantly decreased by the presence of barite on the sand. A 5-mm layer of barite significantly inhibited recruitment of macrofaunal polychaetes and molluscs. The grain size distribution, mineralogy, texture, and organic content of sediments have a profound effect on settlement of planktonic larvae (Thorson, 1957, 1966). Much of the effect of barite on recruitment of benthic invertebrates owed to barium-mediated changes in sediment texture, and not to the chemical toxicity of barite. Barite has a much finer grain size (mean less than 60 μm) than sand. The sand substrate in control aquaria contained no silt-clay fraction, whereas in aquaria containing sand-barite mixtures, the clay-silt fraction was 5.6 to 16.3 percent (Cantelmo et al., 1979). Such changes in sediment characteristics render the sediment more suitable for some species and less suitable for others.

Exposed to a substrate of particulate barite for periods up to 106 days, shrimp *Palaemonetes pugio* ingested the barite (Brannon and Rao, 1979; Conklin et al., 1980). Although this did not affect survival of the shrimp, they were observed to show several sublethal responses. Barite ingestion caused damage to the epithelium of the posterior midgut, possibly by abrasion. The shrimp accumulated barium in the exoskeleton and soft tissues. Barium concentrations in the carapace and other tissues of intermolt shrimp exposed to barite for 21 days were higher than corresponding concentrations in control shrimp. The chemical form and physiological significance of elevated barium concentrations in barite-exposed animals is unknown.

Thompson and Bright (1977) applied separately barite and bentonite clays to small colonies of three reef corals, *Diploria strigosa*, *Montastrea cavernosa*, and *Montastrea annularis*. The surface of the corals was heavily coated, but they were able to clear their surfaces rapidly. *D. strigosa* cleared itself faster than did the other species. Barite and bentonite clays were cleared at about the same rate as natural calcium carbonate sand. During exposure for 29 days to 10 and 100 ppm ferrochrome lignosulfonate in a flowthrough system, polyp retraction in corals *Madracis decactis* was significantly greater than in controls (Thompson, 1980).

Morse et al. (1982) described histological damage to delicate gill ctenidia of sea scallops *Placopecten magellanicus* exposed for up to 20 days to suspensions of 100 ppm or greater of attapulgite clay or to solutions of 500 ppm or greater of ferrochrome lignosulfonate. Ferrochrome lignosulfonate, but not attapulgite, caused a decrease in the rate of the cilia-mediated particle movement across the frontal surface of the gill filaments.

## Hydrocarbons

Because of the low bioavailability of sediment-adsorbed hydrocarbons, most benthic marine animals are able to tolerate higher nominal concentrations of hydrocarbons in sediments than in seawater. Chronic exposure to sediments initially containing 500 to 1,200 ppm crude oil resulted in weight loss and hepatocellular vacuolization in English sole *Paralichthys vetulus* (McCain et al., 1978), reduced condition index and altered tissue-free amino acid ratios in clams *Protothaca staminea* and *Macoma inquinata* (Augenfeld et al., 1980; Roesijadi and Anderson, 1979), and reduced feeding rate in the polychaete *Abarenicola pacifica* (Augenfeld, 1980). Anderson et al. (1978) contaminated natural marine sediments with Prudhoe Bay crude oil, either as a surface layer (5,000 to 6,000 ppm oil initially) or by mixing the oil with the sediments (100 ppm oil initially), and placed the sediments in trays in the intertidal zone of the Washington State coast for 100 days. The oil treatments did not substantially affect recruitment of benthic animals to the sediment trays. When three experimental ecosystems at the Marine Ecosystems Research Laboratory at the University of Rhode Island were dosed with 190 μg/l (ppb) No. 2 fuel oil in the water column for 25 weeks, 109 mg/kg (ppm) petroleum hydrocarbons accumulated in the bottom sediments (Grassle et al., 1980; Oviatt et al., 1982). The macrofauna and meiofauna in the benthos of the oiled tanks declined significantly compared to those in control tanks. The greater effect of No. 2 fuel oil than crude oil on recruitment may owe to the higher concentration of toxic aromatic hydrocarbons in the first or to damage to pelagic larvae caused by the presence of petroleum hydrocarbons in the water column in the experimental ecosystem tanks. More than 5 years after a spill of No. 2 fuel oil near Falmouth, Massachusetts, effects of the oil were still detectable in salt marsh biota where the oil came ashore, probably because the petroleum hydrocarbons persisted in the marsh sediments (Sanders et al., 1980).

### Biocides

Tagatz et al. (1979) studied the effects of three biocides on recruitment of benthic invertebrates to sandy substrates in aquaria. Recruitment of most species to the sand substrate was diminished by two chlorophenol biocides prohibited for ocean disposal, Dowicide G-ST and Surflo B-33. The paraformaldehyde biocide, Aldacide, which is approved for discharge, had no effect on recruitment at concentrations of 14 and 273 μg/l.

## THE TOXICITIES OF USED DRILLING FLUIDS

### Bioassay Procedures

A used drilling fluid, especially a treated one used in a deep hole, is an extremely heterogeneous mixture. It contains water-soluble materials, clay-sized particles of moderate density that settle slowly in seawater, and larger or denser particles that settle rapidly. In addition, montmorillonite and attapulgite clays in fluids flocculate upon contact with seawater, forming larger particles that tend to settle more rapidly than dispersed clay. These fractions tend to separate rapidly when the drilling fluid is added to seawater in a bioassay aquarium, as they do when they are discharged from a drilling rig. Flocculation makes it extremely difficult to design a bioassay protocol in which test organisms are exposed uniformly and reproducibly to a drilling fluid-seawater mixture of known concentration or that at least roughly simulates the kind of exposure an organism might encounter in the vicinity of the drilling-fluid discharge in the field. Because of the complexity of the chemical and physical processes that take place when a used drilling fluid is discharged to the ocean, none of the bioassay protocols used to date are completely satisfactory; however, these protocols are at least consistent with currently accepted bioassay procedures.

The simplest approach has been to add different volumes of whole drilling fluids to a volume of seawater to achieve several concentrations. Test organisms are then exposed to these mixtures, which are aerated, mixed, or left unmixed during the bioassay (Houghton et al., 1980; McLeay, 1976; Tornberg et al., 1980).

Another approach is to evaluate the toxicity of different drilling-fluid fractions or mixtures of drilling fluids and seawater that roughly simulate the likely exposures of organisms in different marine habitats. These bioassay protocols are similar to those recommended for evaluating the environmental impact of dredged material (EPA/COE, 1977). The latter assays have been adopted with minor modifications by EPA for bioassays to comply with NPDES permits for discharging drilling fluids on the mid- and North-Atlantic OCS (Jones and Hulse, 1982). In these EPA bioassay protocols one part drilling fluid is mixed with four parts seawater, and the phases are allowed to separate for one hour. The supernatant is called the suspended particulate phase, and the sedimented fraction the solid phase. A liquid phase is prepared by centrifuging and filtering the suspended particulate phase through a 0.45-μm filter. Other investigators have used an initial dilution of 1 part drilling fluid with 9 parts seawater and a settling period of 20 h (Gerber et al., 1980; Neff et al., 1980, 1981). Protocols using even greater initial dilutions of drilling fluids are being evaluated by investigators associated with the EPA laboratory in Gulf Breeze, Florida.

There is growing evidence that the degree of initial dilution of the drilling fluid has important effects on the composition, and therefore the toxicity, of the three drilling fluid fractions (Neff, unpublished observations). Because of the particulate attraction among clay materials in a drilling fluid slurry, a 4:1 dilution of fluid does not realistically separate into appropriate phases (i.e., suspended particulate and settled solid phases), but rather unnaturally partitions the solid and chemical components of the drilling fluid into the two phases. Field data have shown that drilling fluids are diluted by 1,000 times within a transport period of less than 1 min after discharge (the corresponding distance from the discharge pipe is a function of current, e.g., 4 m at a current velocity of 15 cm/s). This high initial dilution means that the discharge phases occuring in the field would be significantly different than those observed in the protocol of 4:1 dilution. Thus, the method of preparing the bioassay mixture may substantially influence the estimated toxicity of the drilling fluid or drilling-fluid fraction. Bioassay techniques now used also present difficulties in filtering the suspended particulate phase (excess solids stay in suspension) and in the opaqueness of their test solutions (especially of chrome lignosulfonate fluids), which seriously interferes with making bioassay observations on juvenile crustaceans (e.g., mysids). During the course of the bioassay, drilling-fluid solids may accumulate on the bottom of test chambers, creating a viscous zone in which small zooplankton become entrapped. Small copepods, such as *Acartia tonsa*, have been observed mired in layers of settled drilling-fluid solids (Gilbert, 1981), a situation unlikely to occur in nature.

There is some confusion in the literature on the toxicities and biological effects of drilling fluids about the appropriate units to express exposure concentrations and LC50 values. The units most frequently used in the general aquatic toxicology literature are parts per million (ppm), which for liquid or semiliquid solutes in water may be calculated in milligrams or microliters of solute per liter. For solutes with densities near 1.0 kg/l the differences in nominal concentrations calculated in the two ways are not great. The density of used drilling fluids, however, varies from 1.07 to 2.27 kg/l (9 to 19 lb/gal). Large discrepancies can arise in expressing concentrations of drilling fluids in seawater or concentrations of ingredients in drilling fluids as either mg/l or μl/l. A 19-lb fluid might be reported to have a 96-h LC50 value of 1,000 μl fluid/l seawater or one of 2,270 mg fluid/l seawater. The two values are equivalent, but when the numbers are expressed only as "ppm," substantial errors can arise in comparing the toxicities of different fluids. The units expressing exposure concentrations should be clearly defined and standardized. Results are also occasionally reported as ppm phase. If a predilution of 4:1 was used in preparing the phase, then the actual phase concentration has to be multiplied by 0.20 to correct back to whole drilling fluid concentration.

Flow-through exposure systems also have been used for studying long-term and sublethal effects (Conklin et al., 1980; Rubenstein et al., 1980). There is the danger that drilling fluids may fractionate in such systems, and particularly that drilling-fluid solids may accumulate in exposure tanks. No measurements have been made of the variations over time in the concentrations and compositions of drilling fluids in these tanks.

## Acute Lethal Toxicity

The acute lethal toxicities of more than 70 used water-based drilling fluids have been evaluated with 62 species of marine animals from the Atlantic and Pacific Oceans, the Gulf of Mexico, and the Beaufort Sea (Carls and Rice, 1980; Conklin et al., in press; ERCO, 1980; Gerber et al., 1980, 1981; Gilbert, 1981; Houghton et al., 1980; Marine Bioassay Labs, 1982; McLeay, 1976; Neff, 1980; Neff et al., 1980, 1981; Tornberg et al., 1980). Five major animal phyla found in the marine environment were represented by the bioassay organisms, including Chordata (12 species of fish), Arthropoda (30 species of crustaceans), Mollusca (12 species of molluscs), Annelida (6 species of polychaetes), and Echinodermata (1 species of sea urchin). Larvae and other early life stages (considered to be more sensitive than adults to pollutant stress) were also included. The results of these bioassays are summarized in Table 19. Nearly 80 percent of the 400 LC50 values resulting were higher than 10,000 ppm. Two LC50 values were below 100 ppm, both for the copepod Acartia tonsa exposed to heavily treated drilling fluids from Mobile Bay, Alabama (Gilbert, 1981). The estuarine copepod Acartia tonsa and the oceanic copepod Centropages typicus were the most sensitive species tested (EG&G Bionomics 1976b,c; Gilbert, 1981). Other relatively sensitive species included larvae of the dock shrimp Pandalus danae, pink salmon fry Oncorhynchus gorbuscha, larvae of the lobster Homarus americanus, juvenile ocean scallops Placopecten magellanicus and mysid shrimp (Mysidopsis, Neomysis, Acanthomysis [sic Holmeimysis], and Mysis). In most cases, organisms in larval and early juvenile life stages were more sensitive than adults. Molting crustaceans were more sensitive than intermolt animals (Conklin et al., 1980). Crustaceans as a group, and in particular, copepods, mysids, and shrimp, were more sensitive than other major taxa to drilling fluids. This is probably in part because more bioassays were performed with crustaceans in sensitive early life stages than with organisms in other taxonomic groups at these stages. There were no discernible differences in tolerance to drilling fluids among animals from the Atlantic Ocean, Gulf of Mexico, Pacific Ocean, and Beaufort Sea.

Whenever comparisons were made, species were found more sensitive to suspended particulate phase preparations than to liquid phase preparations, indicating that suspended particles or sorbed materials in the drilling fluids contributed substantially to their toxicities. The liquid phase of certain drilling fluids was also toxic. These toxicities may be due to a combination of the chemical toxicity of the liquid phase fluid ingredients and chemicals associated with the particulate phase or the physical toxicity in the form of irritation and damage to delicate gill and other body structures of drilling-fluid particles. Physical abrasion by particles may increase the uptake and therefore the chemical toxicity of soluble components of drilling fluids.

Drilling fluids vary in their toxicities. Information about the types and compositions of drilling fluids used in bioassays is incomplete. Fluids that have been treated heavily with chrome or ferrochrome lignosulfonate, chrome lignosulfonate-bichromate mixtures, surfactants, sulfide scavengers, or diesel fuel are the most toxic. Both the soluble and particulate phases of such fluids are toxic (Conklin et al., in press). Fluids and "spud fluids" (used during initial drilling) have a minimum of additives and toxicities that, in most cases, are not markedly different from that of suspended clay (McFarland and Peddicord, 1980). The soluble fractions of these fluids are usually nontoxic (Neff, 1980; ERCO, 1980).

In summary, LC50s are useful primarily for ranking and comparing the relative toxicities of different chemicals or mixtures and for comparing the sensitivities of different species or life stages to a particular pollutant. The joint IMCO et al. group of experts on the scientific aspects of marine pollution (1969) has used LC50 values to classify different grades or degrees of the acute lethal toxicity of chemicals to marine animals: very toxic chemicals have LC50 values of less than 1 ppm; toxic ones of 1 to 100 ppm; moderately toxic ones of 100 to 1,000 ppm; slightly toxic ones of 1,000 to 10,000 ppm; and practically nontoxic ones of greater than 10,000 ppm. The summary of the results of acute lethal bioassays presented in Table 19 can be interpreted using this classification. Larval, juvenile, and molting crustaceans are more sensitive to drilling fluids than are organisms in most other life stages and of most other species.

## Chronic and Sublethal Effects

Investigations of the chronic and sublethal effects of drilling fluids have been performed with 35 species of marine animals, including 10 species of corals, 5 species of molluscs, 15 species of crustaceans, 1 species of polychaete worm, 2 species of echinoderms, and 2 species of teleost fish. Results of these investigations are summarized in Table 20. The lowest concentrations that elicit a particular response are given. In some experiments, however, this concentration was the lowest concentration tested. Responses to sublethal concentrations of drilling fluids that have been measured include alterations in burrowing behavior and chemosensory responses in lobsters; patterns of embryological or larval development or behavior in several species of shrimp, crab, lobsters, sand dollars, and fish; feeding in larval and adult lobsters and cancer crabs; food assimilation and growth efficiency in opossum shrimp; growth and skeletal deposition in corals, scallops, oysters, and mussels; respiration and nitrogen excretion rates in corals and mussels; byssal thread formation in mussels; tissue enzyme activity in crustaceans; gill histopathology in shrimp and salmon fry; tissue-free amino acid ratios in corals and oysters; and polyp retraction, mucus hypersecretion, ability to clean surfaces, photosynthesis, extrusion of zooanthellae and survival of corals. All the drilling fluids evaluated were chrome or ferrochrome lignosulfonate fluids, the type used most frequently for exploratory drilling on the U.S. OCS. Several of the drilling fluids tested, including the most toxic ones, are known to have contained diesel fuel or other petroleum material. These include several of the fluids from Mobile Bay, Alabama (Conklin et al., in press; Gilbert, 1981 and Jay Field, Florida (Atema et al., 1982; Bookhout et al., 1982; Dodge, 1982; Szmant-Froelich et al., 1982; White et al., 1982) and a medium weight fluid from the Gulf of Mexico (Gerber et al., (1980, 1981); Neff, (1980). Diesel fuels, including No. 2 fuel oil, are known to be quite toxic to marine organisms (Malins, 1977; Neff and Anderson, 1981), and undoubtedly contribute significantly to the toxicity and sublethal effects of those fluids containing them.

Studies of chronic and sublethal effects are often better predictors of the potential environmental impact of a pollutant than are acute lethal bioassays because the first may employ exposure conditions that simulate those organisms might encounter in their natural environment. In most of the investigations summarized in Table 20, this ideal was

TABLE 19 Summary of results of acute lethal bioassays with drilling fluids and marine/estuarine organisms.[a]

| Organism | Number of Species Tested | Number of Fluids Tested[c] | Number of Bioassays Bioassays |
|---|---|---|---|
| Phytoplankton | 1 | 9 | 12 |
| Invertebrates | | | |
| Crustaceans | | | |
| Copepods | 2 | 17 | 39 |
| Isopods | 2 | 4 | 6 |
| Amphipods | 4 | 8 | 19 |
| Mysids[b] | 5 | 18 | 35 |
| Shrimp[b] | 10 | 40 | 76 |
| Crabs[b] | 6 | 18 | 35 |
| Lobsters[b] | 1 | 2 | 7 |
| Molluscs | | | |
| Gastropods | 5 | 5 | 10 |
| Bivalves[b] | 7 | 14 | 33 |
| Echinoderms | | | |
| Sea Urchins[b] | 1 | 2 | 4 |
| Polychaetes | 6 | 14 | 28 |
| Finfish[b] | 12 | 32 | 90 |
| TOTALS | 62 | 72 | 400 |

[a]Most median lethal concentration (LC50) values are based on 96-hour bioassays and results are expressed as parts per million (mg/l or μ l/l) mud added (Based on review of Petrazzuolo, 1981, with data from Carls and Rice, 1980; ERCO, Inc., 1980, Gilbert, 1981, Marine Bioassay Labs, 1982 and Conklin et al., in press, added).

[b]Includes results for embryonic, larval and early life stages.

[c]In many cases, the same drilling fluid was used for bioassays with several species. In a few cases, more than one investigator evaluated the toxicity of a single drilling fluid.

TABLE 19 (continued)

| Organism | Number of LC50 Values (ppm) Not Determinable | 100 | 100-999 | 1,000-9,999 | 10,000-99,999 | 100,000 |
|---|---|---|---|---|---|---|
| Phytoplankton | 5 | 6 | 0 | 7 | 0 | 0 |
| Invertebrates | | | | | | |
| Crustaceans | | | | | | |
| Copepods | 4 | 2 | 11 | 15 | 7 | 0 |
| Isopods | 0 | 0 | 0 | 0 | 1 | 5 |
| Amphipods | 0 | 0 | 0 | 0 | 5 | 14 |
| Mysids[b] | 1 | 0 | 1 | 0 | 21 | 18 |
| Shrimp[b] | 0 | 0 | 12 | 15 | 31 | 18 |
| Crabs[b] | 1 | 0 | 0 | 5 | 16 | 13 |
| Lobsters[b] | 0 | 0 | 0 | 1 | 3 | 3 |
| Molluscs | | | | | | |
| Gastropods | 0 | 0 | 0 | 0 | 2 | 8 |
| Bivalves[b] | 0 | 0 | 0 | 1 | 15 | 17 |
| Echinoderms | | | | | | |
| Sea Urchins[b] | 0 | 0 | 0 | 0 | 1 | 3 |
| Polychaetes | 0 | 0 | 0 | 0 | 9 | 19 |
| Finfish[b] | 0 | 0 | 0 | 3 | 52 | 35 |
| TOTALS | 11 | 2 | 24 | 47 | 163 | 153 |
| Percentage, as a fraction of the total number of drilling fluid bioassays. | 2.8 | 0.50 | 6.0 | 11.75 | 40.75 | 38.25 |

[a]Most median lethal concentration (LC50) values are based on 96-hour bioassays and results are expressed as parts per million (mg/l or μ l/l) mud added (Based on review of Petrazzuolo, 1981, with data from Carts and Rice, 1980; ERCO, Inc., 1980, Gilbert, 1981, Marine Bioassay Labs, 1982 and Conklin et al., in press, added).

[b]Includes results for embryonic, larval and early life stages.

[c]In many cases, the same drilling fluid was used for bioassays with several species. In a few cases, more than one investigator evaluated the toxicity of a single drilling fluid.

TABLE 20 Summary of Investigations of Sublethal and Chronic Effects of Drilling Fluids on Marine Animals (the Lowest Concentration of Drilling Fluid Eliciting a Particular Response)

| Species | Drilling Fluid Type | Exposure Concentration and Duration | Responses | References |
|---|---|---|---|---|
| **Coelenterates (corals)** | | | | |
| *Montastrea cavernosa* *Montastrea annularis* *Diploria strigosa* | Used FeCr-lignosulfonate | 25 ml., 1:1 seawater:fluid | Unable to clear horizontal surfaces | Thompson and Bright, 1977 |
| *Montastrea annularis* | Freshly prepared FeCr-lignosulfonate | 2-4 mm layer applied 4 times at 2.5 h intervals | Decreased growth rate at 6 months | Hudson and Robbin, 1980 |
| *Montastrea annularis* *Porites asteroides* | Used Cr-lignosulfonate, offshore Louisiana | Burial under 10-12 cm for 8 h | All colonies dead after 10 days | Thompson, 1980 |
| *Montastrea annularis* | Used Cr-lignosulfonate, offshore Louisiana | Thin covering | Partial clearing in 26 h, some dead polyps, extruded zooanthellae | Thompson, 1980 |
| *Madracis decactis* | Used Mobil Bay Cr-lignosulfonate with added Cr-lignosulfonate | 100 ppm | Depressed respiration and $NH_3$ excretion rate | Krone and Biggs, 1980 |
| *Porites furcata*, *P. astroides*, *Montastrea annularis*, *Acropora cervicornis*, *Agaricia agaricites* | Used Cr-lignosulfonate, offshore Louisiana | 100 ppm, 96 h | Partial polyp retraction, excess mucus production | Thompson, 1980; Thompson and Bright, 1980; Thompson et al., 1980 |
| *Porites divaricata* | Used Cr-lignosulfonate, offshore Louisiana | 316 ppm, 96 h | Partial polyp retraction, excess mucus production | Thompson, 1980; Thompson and Bright, 1980; Thompson et al., 1980 |
| *Dichocoenia stokesii* | Used Cr-lignosulfonate, offshore Louisiana | 1,000 ppm, 96 h | Partial polyp retraction | Thompson, 1980; Thompson and Bright, 1980; Thompson et al., 1980 |
| *Montastrea annularis* | Used Cr-lignosulfonate with diesel, Jay Field, Fla. | 100 ppm, 6 weeks, flowthrough | 84% reduction in calcification rate, 40% reduction in respiration rate, 26% reduction in photosynthesis, 49% reduction in $NO_3$ and $NH_3$ uptake, inhibition of feeding | Szmant-Froelich et al., 1982 |
| *Montastrea annularis* | Used Cr-lignosulfonate with diesel, Jay Field, Fla. | 100 ppm, 6 weeks, flowthrough | Reduction in skeletal growth rate | Dodge, 1982 |
| *Montastrea annularis* | Used Cr-lignosulfonate with diesel, Jay Field, Fla. | 1-100 ppm, 6 weeks, flowthrough | Growth inhibition, alteration of biochemical pathways and composition, bacterial infection | White et al., 1982 |

TABLE 20 (continued)

| Species | Drilling Fluid Type | Exposure Concentration and Duration | Responses | References |
|---|---|---|---|---|
| **Molluscs** | | | | |
| Pacific oyster Crassostrea gigas | Used medium- and high-weight Cr-lignosulfonate, Gulf of Mexico | 5,000 ppm, 6 weeks, static | Decreased shell growth, decreased condition index | Neff, 1980 |
| Atlantic oyster Crassostrea virginica | Used Cr-lignosulfonate, Mobile Bay, Ala. | 100 ppm, 100 days, flowthrough | Reduced rate of shell regeneration | Rubenstein et al., 1980 |
| C. virginica | Unidentified Cr-lignosulfonate | 4,000 ppm | Altered tissue-free amino acid concentrations and ratios | Powell et al., 1982 |
| Mussel Modiolus modiolus | Used high-weight Cr-lignosulfonate, Cook Inlet, Alaska | 30,000 ppm | Reduced rate of byssus thread formation | Houghton et al., 1980 |
| Mussel Mytilus edulis | Used medium- and high-weight Cr-lignosulfonate, Gulf of Mexico | 33,000 ppm | Decreased filtration rate, increased rate of respiration and $NH_3$ excretion | Gerber et al., 1980 |
| M. edulis | Used medium- and high-weight Cr-lignosulfonate, Gulf of Mexico | 250 ppm[a] | Decreased rate of shell growth | Gerber et al., 1980, 1981 |
| Ocean scallop Placopecten magellanicus (juveniles) | Used medium- and high-weight Cr-lignosulfonate, Gulf of Mexico | 49.4 ppm[a] | Decreased rate of shell growth | Gerber et al., 1981 |
| P. magellanicus (2-day larvae) | Filtered suspension (liquid phase) of used Cr-lignosulfonate, Mobile Bay, Ala., May 15 fluid | 1,000 ppm, 96 h | Significant inhibition of shell formation | Gilbert, 1981 |
| | May 29 fluid | 100 ppm, 96 h | Significant inhibition of shell formation | Gilbert, 1981 |
| P. magellanicus (2-day larvae) | Liquid phase of used Cr-lignosulfonate fluid, Mobile Bay, Ala., September 4 fluid | <100 ppm, 96 h | Signific at inhibition of shell formation | Gilbert, 1981 |
| P. magellanicus (2-day larvae) | Liquid phase of used "Gilsonite" fluid | 3,000 ppm, 96 h | Significant inhibition of shell formation | Gilbert, 1981 |
| P. magellanicus (2-day larvae) | Liquid phase of used low-density lignosulfonate | 10,000 ppm, 96 h | Significant inhibition of shell formation | Gilbert, 1981 |

| | | | | |
|---|---|---|---|---|
| **Crustaceans** | | | | |
| Opossum shrimp <u>Mysidopsis bahia</u> | Used Cr-lignosulfonate Mobile Bay, Ala. | 50 ppm, 42 days, flowthrough | 50% survival from postlarva to adult | Conklin et al., 1980 |
| <u>M</u>. <u>almyra</u> | Liquid phase of used Cr-lignosulfonate, Gulf of Mexico | 10,000 ppm, 7 days | Decreased food assimilation and growth efficiency, reduced growth rate | Carr et al., 1980 |
| Coonstripe shrimp <u>Pandalus hypsinotus</u> (adults) | Used high-weight Cr-lignosulfonate, Cook Inlet, Alaska | 100,000-ppm suspension | Gill histopathology | Houghton et al., 1980 |
| <u>P</u>. <u>hypsinotus</u> (Stage I larvae) | Used FeCr-lignosulfonate Cook Inlet, Alaska | 2,000-ppm suspension, 144 h, 3,250-ppm liquid phase, 144 h | Cessation of swimming by 50% of larvae | Carls and Rice, 1980 |
| Dock shrimp <u>Pandalus danae</u> (Stage I larvae) | Used FeCr-lignosulfonate Cook Inlet, Alaska | 500-ppm suspension, 144 h, 1,050-ppm liquid phase, 144 h | Cessation of swimming by 50% of larvae | Carls and Rice, 1980 |
| Kelp shrimp <u>Eualus suckleyi</u> (Stage I larvae) | Used FeCr-lignosulfonate Cook Inlet, Alaska | 5,000-ppm suspension, 144 h | Cessation of swimming by 50% of larvae | Carls and Rice, 1980 |
| Grass shrimp <u>Palaemonetes pugio</u> larvae | Used medium- and high-weight Cr-lignosulfonate, Gulf of Mexico | 10,000-15,000 ppm liquid phase for duration of larval development | No effect on duration of any intermolt periods or on duration of larval development, significantly increased mortality at molting | Neff, 1980 |
| Sand shrimp <u>Crangon septemspinosa</u> | Used low-weight Cr-lignosulfonate, mid-Atlantic OCS | 33,000-ppm liquid phase, 96 h | Decrease in activity of the enzyme glucose-6-phosphate dehydrogenase in muscle tissue | Gerber et al., 1980 |
| Atlantic Cancer crab <u>Cancer irroratus</u> | Liquid phase of used Cr-lignosulfonate, Mobile Bay, Ala., September 4, 1979 | 100 ppm, 20 days, flowthrough | No effect on survival or molting rate | Gilbert, 1981 |
| <u>C</u>. <u>irroratus</u> (Stage III larvae) | Liquid phase of used Cr-lignosulfonate, Mobile Bay, Ala., September 4, 1979 | 100 ppm, 4 days | Temporary inhibition of feeding | Gilbert, 1981 |
| Cancer crab <u>Cancer borealis</u> | Used medium-weight Cr-lignosulfonate, Gulf of Mexico | 160,000-ppm suspension, 96 h 33,000-ppm liquid phase, 96 h | Increase in activity of enzymes aspartate aminotransferase and glucose-6-phosphate dehydrogenase in heart tissue | Gerber et al., 1981 |

TABLE 20 (continued)

| Species | Drilling Fluid Type | Exposure Concentration and Duration | Responses | References |
|---|---|---|---|---|
| **Crustaceans (continued)** | | | | |
| Green crab *Carcinus maenus* | Used low-weight Cr-lignosulfonate, mid-Atlantic OCS | 33,000-ppm liquid phase, 96 h | Increase in activity of enzymes aspartate aminotransferase and glucose-t-phosphate dehydrogenase in muscle | Gerber et al., 1980 |
| King crab *Paralithoides camschatica* (Stage I larvae) | Used FeCr-lignosulfonate, Cook Inlet, Alaska | 2,800-ppm suspension, 144 h 12,900-ppm liquid phase, 144 h | Cessation of swimming by 50% of larvae | Carls and Rice, 1980 |
| Tanner crab *Chionoecetes* bairdi (Stage I larvae) | Used FeCr-lignosulfonate, Cook Inlet, Alaska | 2,800-ppm liquid phase, 144 h | Cessation of swimming by 50% of larvae | Carls and Rice, 1980 |
| Mud crab *Rhithropanopeus harrisii* larvae | Used low-weight Cr-lignosulfonate, Jay Field, Fla. | 100,000-ppm fluid aqueous fraction and suspended particulate phase, complete larval development | No effect on survival or development rate to first crab stage | Bookhout et al., 1982 |
| Blue crab *Callinectes sapidus* larvae | Used low-weight Cr-lignosulfonate, Jay Field, Fla. | 50,000-ppm fluid aqueous fraction and suspended particulate phase, complete larval development | Significant decrease in survival of megalopa, altered larval behavior | Bookhout et al., 1982 |
| American lobster *Homarus americanus* (adults) | Used low-weight Cr-lignosulfonate, mid-Atlantic OCS | 10,000-ppm liquid phase, 96 h | Increase in activity of the enzyme aspartate aminotransferase and decrease in activity of enzyme glucose-6-phosphate dehydrogenase in heart tissue | Gerber et al., 1980 |
| *H. americanus* (larvae) | Used medium-weight Cr-lignosulfonate, Gulf of Mexico | 2,000-ppm liquid phase | Increase in duration of larval development by 3 days | Gerber et al., 1981 |
| *H. americanus* (adults) | Used medium- and high-weight Cr-lignosulfonate, Mobile Bay, Ala. | 10-ppm suspension, 3-5 min | Decreased chemosensory response of walking leg chemosensors to food cues | Derby and Atema, 1981 |
| *H. americanus* (adults) | Unknown | 1-2 mm layer, 4 days | Inhibition of feeding behavior | Atema et al., 1982 |

| | | | | |
|---|---|---|---|---|
| **Crustaceans (continued)** | | | | |
| *H. americanus* (adults) | Used Cr-lignosulfonate, Mobile, Ala., June 26, 1979 | 7-mm layer, 4 days | No effect on feeding behavior | Atema et al., 1982 |
| *H. americanus* (Stage IV larvae) | Used Cr-lignosulfonate, Jay Field, Fla., July 29, 1980 | 7.7-ppm suspension, 36 days | Partial inhibition of molting, delayed detection of food cues | Atema et al., 1982 |
| *H. americanus* (Stage IV and V larvae) | Used Cr-lignosulfonate fluids, Jay Field, Fla., and Mobile Bay, Ala. | 1-4 mm layer | Delays in burrow construction, altered burrowing behavior | Atema et al., 1982 |
| **Polychaete Worms** | | | | |
| Lugworm *Arenicola cristata* | Used Cr-lignosulfonate fluids, Mobile Bay, Ala. | 10-ppm suspension flowthrough with an accumulation of 4.5 mm at 100 days | 33% mortality | Rubinstein et al., 1980 |
| **Echinoderms** | | | | |
| Sand dollar *Echinarachnius parma* (embryos) | Used Cr-lignosulfonate, MobileBay, Ala., June 26, 1979 | 3,816-ppm[a] suspension, duration of development | Depressed fertilization, delayed development, developmental anomalies | Crawford and Gates, 1981 |
| Bat starfish *Patiria miniata* (embryos) | 13 used Cr-lignosulfonate fluids, Santa Barbara Channel, Calif. | 500-100,000 ppm fluid aqueous fraction, 48 h | Significant decrease in growth rate, increased incidence of developmental abnormalities | Chaffee and Spies, 1982 |
| **Teleost Fish** | | | | |
| Killifish *Fundulus heteroclitus* (embryos) | Used Cr-lignosulfonate, MobileBay, Ala., June 26, 1979 | 3,816-ppm[a] suspension, duration of development | Retarded embryonic development, depressed embryonic heart beat rate | Crawford and Gates, 1981 |
| *F. heteroclitus* (embryos) | Used freshwater Cr-lignosulfonate Gulf of Mexico | 10,000-ppm liquid phase, duration of development | Depressed hatching success, depressed embryonic heart beat rate, developmental anomalies | Sharp et al., 1982 |
| Pink salmon *Oncorhynchus gorbuscha* | Used high-weight Cr-lignosulfonate Cook Inlet, Alaska | 30,000-ppm suspension | Gill histopathology | Houghton et al., 1980 |

[a]Concentrations originally reported as ppm suspended solids, converted here to estimated ppm total fluid added.

not met. Pelagic and some benthic animals were exposed to suspensions or soluble (liquid phase) preparations of drilling fluids continuously for periods of time much longer than they would be in the water column near an exploratory rig. Benthic animals were exposed to layers of whole drilling fluids or to fluids mixed with natural uncontaminated sediments. Unless a drilling fluid is shunted directly to the bottom, it will fractionate as it descends through the water column. Soluble fractions of the fluid not tightly sorbed to clay particles, including the more soluble and toxic aromatic fractions of diesel fuel, may not reach the bottom at all. Lighter clay fractions will be carried farther away than dense barite fractions. Thus, it is unlikely that benthic fauna on the OCS will ever encounter a layer of unfractionated drilling fluid on the bottom. Despite the methodological shortcomings of these studies, however, several of them provide useful insights into the subtle biological responses of marine animals when exposed to sublethal concentrations of drilling fluids. They also suggest the types of responses to look for in field studies of the effects of drilling-fluid discharges on marine communities. It is worth noting, also, that all major taxa have not been treated in a parallel manner in the tests reported in Table 20. Life cycle tests have not been run on indigenous or surrogate fish species.

A difficulty in performing studies of chronic and sublethal responses of marine animals to drilling fluids is that there is no completely satisfactory method for precisely measuring actual exposure concentrations of drilling fluids and their variation over the course of the experiment. Thus, results are presented giving nominal exposure concentrations, based on amount of drilling fluid or fluid fraction added per unit volume of seawater.

Discharges may result in considerable concentrations of suspended solids in the water column (see Table 14), which are rapidly dispersed (see Chapter 3). While the suspended solids themselves may not be toxic, investigators have shown in laboratory studies that solids concentrations may interfere with survival and reproduction of aquatic species (Nimmo et al., 1979; Paffenhofer, 1972; Wilber, 1971). Concentrations of solids that interfered with reproduction in laboratory studies (45 μmg/l)[1] (although differences in reproduction were negligible) are shown in Table 14 to occur as much as 152 m from the point of discharge (assuming a high rate, high volume discharge). However, exposures in the laboratory experiments have been weeks, whereas exposures in the field are minutes.

Biological responses to whole used drilling fluids were recorded at concentrations ranging from 1 to 160,000 ppm and to fluids distributed as a 1-mm to 12-cm layer on natural sediment. In some cases, sublethal responses were observed only at concentrations slightly lower than those that were acutely lethal. For example, the 144-h LC50s of suspended and liquid phase preparations of a used chrome lignosulfonate drilling fluid from Cook Inlet, Alaska, were 1.4 to 3 times higher than

---

[1] Del Nimmo, personal communication, July, 1983.

the concentration of this fluid that in the same period of time caused swimming to cease in 50 percent of Stage I larvae of six species of marine crustaceans (Carls and Rice, 1980). In several other species, however, significant sublethal responses were recorded at concentrations 10 to 100 times lower than the acutely lethal concentrations. These species include the American lobster Homarus americanus and several molluscs, particularly the ocean scallop Placopecten magellanicus (Table 20).

There have been several investigations of the behavioral and physiological responses of reef corals to sublethal concentrations of drilling fluids (Table 20). Exposure to drilling fluid elicits partial or complete polyp retraction in the corals, accompanied in many cases by hypersecretion of mucus. These are defensive reactions that, if they persist for long because of continued pollutant insult, lead to decreased nutrient assimilation and production, altered biochemical composition, depressed respiration and nitrogen excretion, partial or complete inhibition of growth and deposition of calcium carbonate skeleton, bacterial infection, and, eventually, death (Table 20). These responses are elicited by chronic exposure to concentrations of 100 ppm or less, though there are large interspecies differences in sensitivity to drilling fluids. Reef corals are sensitive to drilling fluids, particularly heavily treated ones containing diesel oil.

While this report was in preparation, the previously unpublished results of several investigations became available. These studies measured the acute toxicities and sublethal effects of 11 used drilling fluids obtained from offshore drilling sites in the Gulf of Mexico, which the Petroleum Equipment Suppliers Association supplied to EPA. The mean 96-h LC50 values for bioassays performed with the liquid and suspended particulate phases of drilling fluids and suspended whole fluid preparations for opossum shrimp Mysidopsis bahia were 176,500, 25,145, and 649 μl/l respectively. The mean 96-h LC50 for suspended whole fluid preparations of the 11 drilling fluids for 1-day old larvae of grass shrimp Palaemonetes pugio was 14,516 μl/l. There was a statistically significant inverse relationship between the 96-h LC50 for opossum shrimp and the concentration in the drilling fluids of petroleum hydrocarbons identified as No. 2 fuel oil ($r = -0.73$, $p < 0.05$). Drilling-fluid toxicity was not correlated to concentration of chromium in the fluid ($r = -0.5$, $p > 2$). The drilling fluids contained 100 to 9,430 mg/kg (ppm) petroleum hydrocarbons, and 42 to 1,345 mg/kg total chromium.

When 1 h old embryos of hard shell clams Mercenaria mercenaria were exposed to liquid and suspended particulate phases of the 11 drilling fluids for 48 h, the concentration causing 50-percent inhibition of shell formation ranged from 87 to greater than 3,000 μl/l for the liquid phase and 64 to greater than 3,000 μl/l for the suspended particulate phase. Liquid phase preparations at concentrations as low as 10 to 100 μg/l interfered with fertilization or caused abnormal embryonic development in sand dollars Echinarachnius parma and sea urchins Lytechius variegatus, L. pictus and Strongylocentrotus purpuratus. Reef corals exhibited several sublethal responses following exposure for 24 h to 25 μl/l whole drilling fluid followed by 48 h recovery. These responses included protein loss, changes in the size and concentration ratios of tissue-free amino acids, and depressed calcification rate. The drilling fluids eliciting sublethal responses at the lowest exposure concentrations were those most acutely toxic and containing the highest concentrations of diesel fuel.

## Microcosm Studies

Various types of experimental microcosms have become popular in recent years as links between laboratory experiments and field observations. Microcosms have been used a few times to study the effects of drilling fluids on recruitment of planktonic larvae to benthic communities. In these experiments, drilling fluid is layered on or mixed with the bottom sediment or injected into natural seawater flowing into aquaria (Rubinstein et al., 1980; Tagatz et al., 1978, 1980, 1982). Most of these experiments have tested relatively high concentrations of drilling fluid on or in the sediments (100,000 ppm), which depressed the recruitment of some species. Other species were found in greater numbers in the sediments contaminated with drilling fluids. Certain species of bacteria and microeucaryotes (ciliates, nematodes, etc.) were more abundant in contaminated sediments than in clean ones (Smith et al., 1982). These effects could owe to changes in sediment texture from the presence of drilling fluid, to organic enrichment of sediments, or to the particular chemical compositions of the fluids.

When marine aquaria were supplied with unfiltered natural seawater containing 50 μl/l (ppm) of used chrome lignosulfonate drilling fluid for 8 weeks, the numbers of tunicates, molluscs, and annelids settling in the sandy substrate or on the walls of the aquaria were significantly decreased compared to those settling in control aquaria (Tagatz et al., 1982). Differences in community structure in control and experimental aquaria receiving 50 ppm drilling fluid were indicated by a decrease in species abundance by Spearman's measure of rank correlation and an increase in species diversity as measured by the Shannon-Weaver index. These differences could have owed to the physical or chemical effects of suspended drilling fluids on survival or settlement of planktonic larvae or to the accumulation of drilling fluids in the sediments over time (which was noted but not quantified) altering sediment texture.

## Field Studies

Table 21 provides summary information on the few field investigations that have been conducted of the environmental fate and effects of drilling fluids and cuttinggs discharged to the marine environment. These studies corroborate predictions derived from laboratory studies. The effects of drilling-fluid discharges to marine ecosystems, where detected, are localized to an area around and downcurrent of the discharge and to the benthos.

Gettleson (1978) monitored the condition of reef corals on the East Flower Garden Bank off the Texas-Louisiana coast before, during, and

TABLE 21 Summary of Major Field Investigations of the Environmental Fate and Effects of Drilling Fluids and Cuttings Discharged to the Environment

| Location | Objectives | Physical Characteristics | Results | References |
|---|---|---|---|---|
| East Flower Garden Bank, NW Gulf of Mexico | Fate of drilling fluids shunted to 10 m above bottom; effects on coral reef 2,100 meters away | Drilling at 129 m water depth; coral zone at 20-50 meters & NW of drill site; bottom currents toward WSW drill site | Drill fluids and cuttings distributed to 1,000 m from discharge; no impact on coral zone | Gettleson, 1978 |
| Palawan Island, Phillippines | Effects of drilling discharges on coral reefs | Drilling directly on reef at 26 m 2 wells drilled 3 m apart; 3 cm/s currents to the north | 70-90% reduction in some spp. of living corals within 115 x 85 m area; epifauna associated with corals affected to 40 m | Hudson et al., 1982 |
| Lower Cook Inlet, AK | Fate of drilling discharges and effects on benthic communities | Drilling at 62 m water depth, 4.6-5.3 m tides, mean maximal tide currents 42-104 cm/s between bottom and surface | Little accumulation of mud & cuttings on bottom; no effects on benthos attributable to discharges | Dames & Moore, 1978<br>Houghton et al., 1980<br>Lees & Houghton, 1980 |
| NJ 18-3 Block 684, Mid-Atlantic OCS | Fate of drilling discharges; effects on benthic community; bioaccumulation of metals | Drilling at 120 m water depth; bottom currents < 10 cm/s 62% of time, sediments 20% silt/clay | Visible cuttings pile 150 m diameter; elevated Ba in sediments to 1.6 km; abundance of predatory demersal spp. increased; large decrease in abundance of benthic infauna near rig with some bioaccumulation of Ba and possibly Cr by benthic infauna | EG&G Environmental Consultants, 1982 |
| Georges Bank, North-Atlantic OCS | Fate of drilling discharges; effects on benthic community; bioaccumulation of metals | Rigs at 80 and 140 m monitored; residual bottom current 3.5 cm/s; Frequent severe storms; sediments < 1% silt/clay | Evidence of cuttings within 200 m of rigs; elevated Ba in bulk sediments to 2 km; no effects on benthos attributable to drilling; no bioaccumulation | Battelle/W.H.O.I., 1983<br>Bothner et al., 1982<br>Payne, et al., 1982 |
| U.S. Beaufort Sea, AK | Effects of above-ice and below-ice disposal of drilling mud and cuttings on benthic communities; bioavailability of metals | Water depth 5-8 m; ice cover most of year with bottom scour in shallower areas | 0.5-6 cm fluid and cuttings on bottom but carried away quickly; no effects attributable to discharges on benthos; possible uptake of Ba by macroalgae and Cu by amphiphods | Northern Technical Services, 1981 |
| Offshore Southern California | Effects of drilling discharges on fouling community on pontoons of semisubmersible rig | Platform on site 2-3 years, sampling in August | Surfaces within 10 m of discharge had different fouling community, attributed to drilling fluid accumulation | Benesch et al., 1980 |
| Canadian Beaufort Sea | Metals from drilling discharges in sediments and benthos | Drilling from artificial island; rapid seasonal erosion and ice scour | Elevated levels of Hg, Pb, Zn, Cd, As, and Cr in sediments near discharge with elevated Hg to 1,800 m; no correlation between metals in sediments and biota | Crippen et al., 1980 |
| NW Gulf of Mexico | Distribution of metals in sediments and biota in oil production fields | Shallow water, high suspended sediment load | Decreasing concentration gradients of Ba, Cd, Cr, Cu, Pb, and Zn in sediments around some rigs. Metals not elevated in commercial species of shrimp and fish | Tillery and Thomas, 1980 |

after the drilling of an exploratory well approximately 2,100 m southeast of the reef. Discharges were shunted to 10 m off the bottom. Although some of the discharged fluid and cuttings were distributed by currents to a distance greater than 1,000 m from the rig, none could be detected in the coral reef zone, which was shallower than the depth of discharge.

Hudson et al. (1982) found little or no suppression of growth in the coral *Porites lutea* from drilling-fluid discharges made in exploratory drilling near a coral reef off Palawan Island, Philippines. Living foliose, branching, and plate-like corals were reduced by 70 to 90 percent, however, in an area 115 by 85 m around the wellheads, possibly because of the smothering or toxic effects of these discharges. Communities of small organisms living in crevices and cavities in and among the coral heads (coelobites) were severely disturbed within 40 m of the wellheads (Choi, 1982). Minor changes in coelobite community structure were observed up to 100 m from the wellhead. Animals living on the surface of the reef were less affected.

Lees and Houghton (1980) studied benthic communities in the vicinity of the C.O.S.T. well in Lower Cook Inlet, Alaska, before, during, and after the drilling operation. Changes in benthic communities were seen near the drilling platform during the course of the study. None could be unequivocably attributed to the drilling operations, however, because of irregularities in faunal distribution, probably owing to differences in successional stages among the areas sampled and the failure to resample control sites. They concluded that, in the very high energy environment of Lower Cook Inlet, the rate of accumulation of drilling fluids and cuttings on the bottom was not sufficient to affect measurably benthic populations. Although populations of an opportunistic species of polychaete, *Spiophanes bombyx*, may have increased after drilling, such resistence is characteristic of dynamic environments. In a related study of the same drilling rig, Houghton et al. (1980) placed pink salmon fry, shrimp, and hermit crabs in live boxes at 100, 200, and 1,000 m downcurrent from the drilling-fluid discharge. In an observation made after 4 days, no mortalities could be attributed to the fluid's discharge plume.

Detailed studies have been performed on the shortand long-term effects of drilling fluids and cuttings on benthic communities around an exploratory drilling platform in New Jersey 18-3 Block 684 on the mid-Atlantic OCS off Atlantic City, New Jersey (EG&G Environmental Consultants, 1982; Gillmor et al., 1981, 1982; Maurer et al., 1981; Menzie et al., 1980). A zone approximately 150 m in diameter of visible accumulation from drilling discharges (primarily from drill cuttings) was observed in the immediate vicinity of the well site, while elevated levels of clays were detected up to 800 m southwest of the site immediately after drilling ceased (during a first post-drilling survey 2 weeks later). A side-scan sonar survey 1 year after drilling ceased revealed scour marks left by anchor chains and depressions left by the anchors. Drill cuttings and debris had accumulated heavily in an area about 40 to 50 m in diameter immediately south of the well site. The height of the cuttings pile was estimated to be less than 1 m. During the second postdrilling survey, elevated levels of clay were not detected southwest of the drill site. In both postdrilling surveys, concentrations of barium in the upper 3 cm of sediments were elevated (up to 3,477 ppm in the first survey and 2,144 ppm in the second survey, compared to 148 to 246 ppm before drilling) near the rig site and decreasing with distance from the rig. The concentration of barium was elevated in sediments up to 1.6 km from the drill site. Neither the concentration of chromium nor of several other metals was elevated in sediments near the rig following drilling.

The abundance of hake (Urophycis chuss), cancer crabs (Cancer spp.), and starfish (Astropecten americanus) increased between the predrilling and first postdrilling surveys in the immediate vicinity and to the south of the well site. These animals may have been attracted by the increased microrelief of the accumulated cuttings or by clumps of mussels Mytilus edulis that had fallen off the drilling rig or anchor chains. Within about 150 m of the discharge, sessile benthic animals such as sea pens Stylatula elegans were subject to burial by drill cuttings. The second postdrilling survey found sea pens completely absent from the main cuttings pile, although they were observed among patches of cuttings away from it. One year after drilling, hake and cancer crabs were no longer concentrated near the rig site, and starfish had a patchy distribution throughout the area.

Before drilling, the abundance of benthic macrofaunal in the vicinity of the rig site was greater than that at a nearby BLM benchmark station (8,011 animals/$m^2$ versus 3,064 animals/$m^2$). The abundance of benthic macroinfauna at the rig site dropped to 1,729 animals/$m^2$ immediately after drilling, and then rose to 2,638 animals/$m^2$ one year later. These changes in abundance were the same for the four major taxonomic groups (polychaetes, echinoderms, crustaceans, and molluscs). Polychaetes predominated in the macroinfauna at the study site during all three survey periods. Their relative abundance, however, dropped from 78 percent in the predrilling survey to 70 percent in the first postdrilling survey and to 66 percent in the second postdrilling survey, compared to 70 percent at the nearby BLM benchmark station. Molluscs were the only group to return to their original abundance at the site within 1 year after drilling.

The abundance of the brittle star Amphioplus macilentus substantially decreased within 100 m of the rig site and remained decreased at the time of the second postdrilling survey. The abundance of small brittle stars (of disc diameter less than 1.5 mm) decreased more than that of larger specimens. The number of polychaetes measured during the first postdrilling survey was significantly lower at stations near the rig site that had elevated levels of clay (from drill cuttings) compared with nearby stations that show no elevation in sediment clay concentration between predrilling and postdrilling surveys. The composition of polychaete feeding guilds, however, was similar in all three surveys (Maurer et al., 1981). With the exception of these cases, benthic macrofauna decreased in abundance similarly between the predrilling survey and the two postdrilling surveys for the major species within all taxnomic groups. With the exception of a few stations less than 100 m southwest (downcurrent) of the drill site that had markedly reduced benthic fauna during the first postdrilling survey, there was no relationship between direction, distance from the rig site (out to 3.2 km), or sediment barium concentration and the extent of decrease in abundance of any major taxonomic group or major species.

Unfortunately, there were no control stations sufficiently far from the rig site to ensure that they were not affected and thus that they provided true reference points to evaluate the three benthic samplings. Data from more distant stations might establish better how much changes in benthic fauna resulted from drilling or from other factors. Stations farthest from the rig site and considered beyond the influence of drilling discharges (Stations 55, 56, and 58) showed the same patterns of faunal change as did stations near the rig site. The composition and abundances of benthic fauna observed in the two postdrilling surveys were more like those observed in the earlier BLM benchmark program in the area, particularly from BLM Station A3 (Boesch, 1979), than like those observed in the predrilling survey. Because of natural temporal variability, the predrilling survey may not have provided a suitable baseline (at least as far as macrobenthos abundance levels are concerned) to evaluate the results of the postdrilling surveys. Natural temporal variability is the probable cause for the large, area-wide changes in macrobenthic abundance that was observed between surveys. This premise is supported by considering that an exploratory well was drilled approximately 2.8 km north (upcurrent) of the monitored well site shortly before the predrilling survey. If area-wide impacts occurred as a result of drilling this well, they should have influenced the stations north of the test well; however, no differences in composition and abundance were observed between these stations and stations south of the test well. The previously-drilled well was drilled by the same operator using the same drilling fluid company and program and drilled through similar formations as the test well. Water depth, bottom topography and currents at both sites are similar. Thus similar distributions of drilling discharges around each well would be expected. If natural variability is ignored, pre-drilling macrofaunal abundance appears to be elevated (with respect to BLM benchmark data) even though some of the stations could have been exposed to impacts from the previous well. However, data from the monitored well suggest that macrofaunal abundances decreased upon exposure to drilling discharges. This contradiction strengthens the arguments that, except for those stations in the immediate well-site area, the observed decrease in macrofaunal abundance between pre- and post-drilling surveys resulted from natural temporal variability.

Species richness (number of species per 0.2 $m^2$) at the rig site dropped from 70 $\pm$ 7 in the predrilling survey to 38 $\pm$ 10 immediately after drilling and then rose again to 53 $\pm$ 8 one year later. Shannon diversity (H') and evenness (J') showed only very small changes between the predrilling and the two postdrilling surveys. Diversity decreased slightly, probably in part because of increased evenness, which was observed in the postdrilling surveys. These changes in species richness, diversity, and evenness were similar at stations near the well site and at the three stations considered to be beyond the influence of drilling discharges.

The authors concluded that the physical and biological effects of exploratory drilling discharges on the benthic environment of a low-energy area of the mid-Atlantic OCS persisted for at least 1 year after drilling activities ceased. To the extent that the decreased abundance and species richness of benthic macrofauna around the rig site immediately after drilling resulted from drilling discharges, there was evidence of recovery during the year after drilling ceased.

A similar investigation is being performed for the Minerals Management Service (formerly the Bureau of Land Management) on Georges Bank, southeast of the Massachusetts coast (Battelle/Woods Hole Oceanographic Institution, 1983; Bothner et al., 1982; Payne et al., 1982). Small amounts of cuttings were detected in bottom sediments within about 200 m of the exploratory rigs in Blocks 312 (94 m water depth) and 410 (137 m water depth) following drilling (Bothner et al., 1982). No pile of cuttings was visible in any bottom photograph. Barium concentration increased in the top centimeter of bulk sediments between predrilling and postdrilling surveys up to 3.5 times (from 32 to 110 ppm) within 200 m of the rig in Block 410. A smaller increment in sediment barium concentration was observed in the upper centimeter of sediments collected from within 200 m of the rig site in Block 312. Elevated levels of barium, but not chromium, were detected in bulk sediment samples up to 2 km from both drill sites. The silt-clay fraction of the sediments, representing about 1 percent of the total, contained elevated concentrations of barium and chromium at stations up to 6 km downcurrent of the Block 312 drill site after 6 months of drilling.

During the first year of the monitoring program, benthic samples were collected four times on a seasonal basis (in July and November 1981, and in February and May 1982) from 47 sampling stations upcurrent, in the vicinity, and downcurrent of the lease blocks. Drilling began at the two rig sites in Blocks 312 and 410 in December and July 1981 respectively. Drilling was observed to have little impact on the abundant and diverse benthic macroinfauna during the first year of monitoring (Battelle/Woods Hole Oceanographic Institution, 1983). In Block 312, where drilling started shortly after the second survey, there was a change in the abundance of several species at some stations within 2 km of the rig where Bothner et al. (1982) showed that barium (and by inference the solid components of drilling fluids) accumulated between the first and fourth surveys. In February, shortly after drilling started, some species increased in abundance at stations closest to the rig and declined at stations farther away. The abundance of the corophiid amphipod *Erichthonius rubicornis* showed a marked decline in February at some stations. Barium was not observed to accumulate at these stations until May. Thus it is doubtful that the population changes observed resulted directly from the accumulation of discharged drilling fluids on the bottom, since the discharges accumulated after the amphipod population had declined. However, the distribution and abundance of *E. rubicornis*, an epifaunal suspension feeder, and of certain other species around the rig may have been influenced by the accumulation on the bottom of drill cuttings, most of which are discharged during the drilling of the shallow portion of the hole early in drilling (Ayers et al., 1980a). Severe winter storms in February 1982, however, caused substantial sediment resuspension and bottom scour at these stations, as documented by bottom photography. Changes in sediment texture resulting from the storms probably were a major cause of the benthic infaunal changes seen near the rig in February. Most of the macrofaunal species that declined in abundance near the rig site in February substantially increased in abundance in May. Thus, any effects of drilling discharges were apparently of short duration.

The much milder effects of exploratory drilling on the benthos of Georges Bank than those on the mid-Atlantic OCS probably result in large part from the difference in the amounts of drilling fluids and cuttings accumulating on the bottom at the two sites. The lower energy environment of the mid-Atlantic OCS allowed more drilling fluids and cuttings to accumulate on the bottom than did the higher energy environment of Georges Bank.

Northern Technical Services (1981) investigated the effects of above-ice and below-ice disposal of drilling fluids and cuttings on the nearshore benthos of the U.S. Beaufort Sea. Approximately 2.6 x $10^4$ l of drilling effluents were discharged below the ice at shallow-water (5.5-m) and deep-water (8.2-m) test locations near the Reindeer Island Stratigraphic Test Well site approximately 15 km north of Prudhoe Bay, Alaska. In addition, approximately 3.5 x $10^5$ l of drilling effluents were discharged on the ice in 6.7 m of water. Reference sites were located nearby in 4.9 and 7.67 m of water. Four days after the test discharge at the deep water site, a layer of drilling fluid and cuttings of 5 to 6 cm was observed on the bottom under the discharge point. About 3 m east of the site the estimated depth of the layer was about 0.5 cm. At the shallow water site the maximum accumulation of drilling fluid and cuttings was about 1 to 2 cm.

In order of relative abundance, the benthic fauna of the study area included polychaetes, molluscs, and crustaceans. The experimental and reference stations in shallow water (5 m) and in deep water (8 m) differed significantly in infaunal abundance, diversity, species richness, evenness, and biomass. The experimental discharges took place in late April and early May 1979. Analysis indicated that the abundance of some species changed at the experimental and reference sites between May and August. The changes probably were due to seasonal effects. At the disposal site above ice, the numbers of polychaetes and harpacticoid copepods were significantly fewer than at the nearby deep-water reference site in August 1979 and January 1980. Grain size and trace metal analyses of bottom sediments from the two sites indicated that drilling effluents did not remain for long at the disposal site. The authors attributed the differences in polychaete and harpacitcoid abundances at reference and above-ice disposal sites to natural differences in ambient physical conditions (mainly sediment grain size) at the two sites.

Amphipods (*Onisimus* species and *Boeckosimus* species), placed in live boxes on the bottom or at mid-depth 3 to 12 m from the discharge points for 4 to 89 days suffered few mortalities. Trays containing clams (*Astarte* species and *Liocyma fluctuosa*) were deployed for up to 89 days on the bottom at the deep-water reference site and the above-ice experimental discharge site. After 4 days, 1 to 2 clams were dead in both reference and experimental trays. After 87 to 89 days, 7 clams (26 percent) were dead in the experimental tray and 9 were missing, compared to 1 dead in the reference tray. The experimental tray had also been disturbed, however, which could have contributed to the mortalities observed.

Benech et al. (1980) studied fouling communities on submerged pontoons of a semisubmersible drilling rig off southern California. The horizontal pontoon surfaces within 10 m downcurrent of the discharge pipe where solids accumulated had different fouling communities than pontoon surfaces where these solids did not settle. Differences were attributed primarily to sedimentation of the drilling fluids and cuttings. Sediment-intolerant species disappeared and sediment-tolerant species became more abundant on the fluid-exposed pontoons.

In summary, the effects of drilling fluids and cuttings on benthic and fouling communities is related to the amount of material accumulating on the substrate, which in turn is related to current speed and related hydrographic factors. In a high-energy environment, fluids and cuttings do not accumulate and have not been observed to affect the benthos. In low-energy environments, more material accumulates, and in the vicinity of the drill site the abundance of certain benthic species is reduced as a result of burial, the species' incompatibility with clay, or the chemical toxicities of the components of drilling fluids or cuttings.

## BIOAVAILABILITY

### Hydrocarbons

Highly aromatic diesel fuels (containing 30 to 40 percent aromatics) such as No. 2 diesel fuel are among the most toxic petroleum products to marine organisms. Most of the petroleum hydrocarbons in a used diesel-treated drilling fluid probably will be sorbed to the bentonite clay fraction of the fluid and be incorporated in the sediments. Petroleum hydrocarbons sorbed to organic or inorganic particles generally are less bioavailable to marine organisms than hydrocarbons in solution or dispersed in the water column (Augenfeld et al., 1982; McCain et al., 1978; Roesijadi et al., 1978a,b; Rossi, 1977; Lyes, 1979; Neff, 1979). The bioconcentration factor (concentration in tissue/concentration in sediments) for petroleum hydrocarbon uptake from sediments and detritus by marine animals usually falls in the range of 1 to 2. Augenfeld et al. (1982) reported maximum bioaccumulation factors of 7.9 and 11.6 for phenanthrene and chrysene respectively by the clam *Macoma inquinata* from sediments. Although particle-sorbed petroleum hydrocarbons are less bioavailable than hydrocarbons in solution, there could be sufficient uptake of hydrocarbons from drilling fluids to contribute significantly to the toxicity of those fluids that contain diesel oil. There have been no published laboratory investigations to date of the uptake of petroleum hydrocarbons from diesel-treated drilling fluids.

### Heavy Metals

Metals commonly found in drilling fluids are barium, chromium, cadmium, copper, iron, mercury, lead, and zinc (Table 22). Compounds containing

TABLE 22 Trace Metal Concentrations in Drilling Fluids From Different Sources[a]

| Drilling Fluid | Ba | Cr | Cd | Cu | Fe | Hg | Pb | Zn | Others | References |
|---|---|---|---|---|---|---|---|---|---|---|
| 48 Canadian Arctic fluids | ND | 0.1-909 | ND | 0.05-250 | 0.002-9,250 | ND | ND | 0.06-1,700 | | Siferd, 1976 |
| 3 Barite CLS fluids | ND | ND | 0.16-54.4 | 6.4-307 307 | ND | 0.2-10.4 | 0.4-4,226 | 6.6-12,270 | 3.8-19.9 Ni | Nelson et al., 1980 |
| 2 Mid-Atlantic CLS fluids | 229,100-303,700 | 1,112-1,159 | 0.6-0.8 | 5.8-7.7 | ND | $<0.05$ | 102.6-218.5 | 36.0-48.4 | 1.8-2.3 As<br>13.5-17.0 Ni<br>22.7-28.0 V | Ayers et al., 1980a |
| 3 Mid-Atlantic CLS fluids | 823-19,300 | 57-90 | 2 | ND | ND | 1-2.8 | 10-241 | 101-197 | 20-33 | EG&G, 1980 |
| Baltimore Canyon CLS fluid | 202,000 | 850 | ND | 20 | 19,000 | ND | ND | ND | | Liss et al., 1980 |
| Gulf of Mexico CLS fluid | 449,000 | 378 | ND | ND | ND | ND | ND | ND | 10,800 Al | Ayers et al., 1980b |
| Gulf of Mexico CLS fluid | 133,000 | 200 | ND | 280 | 16,000 | ND | ND | ND | | Liss et al., 1980 |
| Gulf of Mexico Spud Fluid[b] | ND | 51 | 0.51 | ND | ND | ND | ND | ND | | Page et al., 1980 |
| Gulf of Mexico high-density CLS fluid[b] | ND | 257 | 0.78 | ND | ND | ND | 1.3 | ND | | Page et al., 1980 |
| Gulf of Mexico mid-density CLS fluid[b] | ND | 396 | 1.70 | ND | ND | ND | 5.0 | ND | | Page et al., 1980 |
| Gulf of Mexico low-density CLS fluid[b] | ND | 596 | 1.18 | ND | ND | ND | ND | ND | | Page et al., 1980 |
| Gulf of Mexico seawater CLS fluid | ND | 485.2 | 3.0 | 48.2 | ND | ND | 179.4 | 251.4 | | McCulloch et al., 1980 |
| Gulf of Mexico spud fluid[b] | ND | 10.9 | 3.5 | 30.2 | ND | ND | 134.2 | 297.3 | | McCulloch et al., 1980 |
| Gulf of Mexico high-density CLS fluid[b] | ND | 229.9 | 10.9 | 118.8 | ND | ND | 209.5 | 274.5 | | McCulloch et al., 1980 |
| 2 CMC/gel fluids Alaska[c] | 4,400-6,240 | 28-63 | 0.5-0.6 | 6.4-10.4 | ND | 0.017-0.031 | 2.4-12.8 | 42-64 | | Tornberg et al., 1980 |
| XC polymer fluids Alaska (20)[c] | 720-1,120 | 66-176 | 0.5-1.5 | 10-16 | ND | 0.015-0.070 | 5.6-56 | 49-110 | | Tornberg et al., 1980 |
| XC polymer/unical fluids Alaska (6)[c] | ND | 56-125 | ND | 2.8-17.0 | ND | 0.028-0.217 | 9-117 | 198-397 | | Tornberg et al., 1980 |
| CLS fluids Alaska (4)[c] | 800-7,640 | 121-172 | 0.5 | 10-12 | ND | 0.03-0.07 | 16.4-56.0 | 49-56 | | Tornberg et al., 1980 |
| Gulf of Mexico CLS fluid | 90,000 | 500 | ND | 43 | 27,000 | ND | 91 | 370 | 400 Mn | Trefry et al., 1981 |
| Mobil Bay treated CLS fluid | ND | 5,960 | ND | 47 | 10,100 | ND | 22 | 540 | 290 Mn | Trefry et al., 1981 |

NOTE: ND, not determined; CLS, chrome lignosulfonate; CMC, carboxymethylcellulose.

[a]Concentrations are in mg/kg dry weight (ppm).
[b]Fluids also analyzed by Page et al. (1980) and McCulloch et al. (1980).
[c]Concentrations given on a wet-dry basis.

barium, chromium, lead, and zinc are intentionally added to drilling fluids to serve different functions. Other metals the fluids contain are trace contaminants of barite and bentonite clay and formation solids (Kramer et al., 1980; MacDonald, 1982). Elevated concentrations of barium, and occasionally chromium, zinc, cadmium, and lead, presumably derived in part from discharged drilling fluids, have been reported in the water, bottom sediments, or both in the immediate vicinity of offshore exploratory wells (Crippen et al., 1980; Ecomar, 1978; EG&G Environmental Consultants, 1982; EG&G Environmental Consultants, 1982 Gettleson and Laird, 1980; Meek and Ray, 1980; Tillery and Thomas, 1980; Trocine et al., 1981; Wheeler et al., 1980;). The important question relating to these metals is whether marine animals can accumulate them in their tissues from the water or sediment to the extent that the metals are toxic to the animals themselves or to animals at higher trophic levels, including, for example, human consumers of fishery products.

### Laboratory Studies

There have been a number of laboratory investigations of the bioaccumulation of some metals in drilling fluids or drilling-fluid ingredients (Brannon and Rao, 1979; Carr et al., 1982; Espey Huston & Associates, 1981; Gerber et al., 1981; Liss et al., 1980; McCulloch et al., 1980; Page et al., 1980; and Rubinstein et al., 1980). They show that some heavy metals in used drilling fluids are bioavailable to marine animals. Statistically significant bioaccumulation of chromium and barium may occur, despite the very low solubility of barium sulfate in seawater. Liss et al. (1980) have shown that higher concentrations of chromium and barium than predicted are present in filtrates of seawater suspensions of drilling fluids; this may be the fraction accumulated by marine animals. Much of the lead, zinc, and possibly cadmium is in particulate form and associated with pipe dope (usually high in lead and zinc) and in the clay or barite fractions of the fluids (Kramer et al., 1980; MacDonald, 1982; McCulloch et al., 1980).

### Field Studies

Several metals in drilling fluids, particularly barium, tend to accumulate in bottom sediments in the immediate vicinity and downcurrent of the drilling rig, where they may persist indefinitely (Boothe and Presley, 1983; Crippen et al., 1980; EG&G Environmental Consultants, 1982; Gettleson and Laird, 1980; Meek and Ray, 1980; Tillery and Thomas, 1980; Trocine et al., 1981; Wheeler et al., 1980). The question of the bioavailability of these sedimented metals to benthic marine animals has been explored by Crippen et al. (1980), and Tillery and Thomas (1980).

Changes were reported in concentrations of several metals in sediments and benthic invertebrates in the vicinity of an offshore exploratory rig in the Baltimore Canyon off New Jersey before and after drilling (EG&G Environmental Consultants, 1982). Only the elevations in barium concentration in the postdrilling sediment samples could be attributed to the drilling-fluid discharges. Concentrations of chromium in sediments from the postdrilling surveys were within the range of values obtained for sediments from the predrilling survey and from BLM stations A2 and A3 near the drill site sampled on five surveys prior to exploratory drilling. Other investigators have identified barium as the metal most enriched in bottom sediments around drilling-fluid discharges (Bothner et al., 1982; Chow and Snyder, 1980; Gettleson and Laird, 1980; Wheeler et al., 1980). This is not surprising given the high density and low solubility of barite and the large amounts of it used in most fluids when drilling deep.

Some samples of mixed-species assemblages of brittle stars, molluscs, and polychaetes collected during the first and second post-drilling surveys, at approximately 2 weeks and 1 year after drilling ceased, had significantly elevated concentrations of barium and chromium compared with animals collected in the predrilling survey nearly 1 year before drilling started (EG&G Environmental Consultants, 1982). The reported increase in mercury concentration in tissues of animals from the first postdrilling survey (Mariani et al., 1980) was later found to be in error (EG&G Environmental Consultants, 1982). Recalculation of the range of mercury concentrations in molluscs, brittle stars and polychaetes revealed no statistically significant increase in mercury concentration between biota sampled before and after drilling.

In both postdrilling surveys the concentrations of barium in tissues of molluscs from the immediate vicinity of the drill site were within the range observed during the predrilling survey. Barium concentrations in tissues of polychaete worms and brittle stars from the vicinity of the drill site were significantly higher in samples from the first postdrilling survey than in those collected before drilling started. Mean barium concentrations in polychaetes and brittle stars were 24 and 15 ppm before drilling, and 88 and 218 ppm during the first postdrilling survey. One year after drilling ceased, barium concentrations in all but a few polychaete and brittle star samples had returned to those observed prior to drilling. Concentrations of chromium were elevated in tissues of polychaetes during the first postdrilling survey, and in tissues of molluscs, polychaetes, and brittle stars during the second postdrilling survey. Concentrations of barium and chromium in the tissues of benthic organisms were not correlated with the concentration gradients of these metals in bottom sediments.

Payne et al. (1982) could find no indication of any increase in the concentration of barium, chromium, or several other metals in the tissues of bivalve molluscs *Arctica islandica* or of demersal fish near exploratory drilling on Georges Bank. A few mollusc samples collected in February and May 1982 contained slightly elevated levels of petroleum hydrocarbons, but the source of these could not be identified.

Concentrations of several metals were measured in tissues of macro-invertebrates and macroalgae from the bottom at a reference and above-ice drilling-fluid disposal site in the Beaufort Sea 8 and 12 months after an experimental discharge (Northern Technical Services, 1981). Most metals were present in higher concentrations in organisms from the reference than from the experimental site. The concentration of barium was found to be elevated in polychaete tubes and macroalgae (_Eunephyta rubriformis_) from the experimental site, but this concentration was analyzed by atomic absorption spectrometry, so the results may not be reliable. The macroalgae also had slightly elevated levels of chromium (3.86 compared to 1.54 μg/g dry weight at experimental and reference sites). Amphipods maintained in live boxes for 89 days at the experimental site contained slightly elevated levels of copper (114 compared to 89.5 μg/g dry weight at experimental and reference sites). Concentrations of other metals analyzed (chromium, lead and zinc) were similar in both experimental and control groups.

Crippen et al. (1980) measured the concentrations of several metals in sediments, drilling fluids, and benthic animals from a drilling site in the Beaufort Sea. Mercury, lead, zinc, cadmium, and arsenic were present at higher concentrations in the drilling fluid than in the surface sediment. Some of these metals were associated with an impure grade of barite used to formulate the drilling fluid and probably were in the form of insoluble metallic sulfides (Macdonald, 1982). Metal levels in the sediment near the discharge site were not significantly correlated to those found in nearby benthic infaunal organisms.

Tillery and Thomas (1980) reviewed several investigations of the distribution of heavy metals in sediments and biota in oil production fields in the northwest Gulf of Mexico and found that the concentration gradients of barium, cadmium, chromium, copper, lead, and zinc in surficial sediments decreased with distance from some platforms. Trace metal concentrations in muscle tissues of four commercially important species (brown shrimp _Penaeus aztecus_, Atlantic croaker _Micropogon undulatus_, sheepshead _Archosargus probatocephalus_, and spadefish _Chaetodipterus faber_) generally were not significantly higher in animals from the vicinity of oil production fields than in animals from other regions. They found, however, that such metal concentrations were not determined for other tissues, some of which are more likely than muscle to accumulate metals.

The results of the limited field studies tend to corroborate the results of laboratory studies. The accumulation in organisms of heavy metals from sedimented drilling fluids is low. Most of the metals of concern are originally associated with the barite and bentonite clay fractions of the drilling fluid (Crippen et al., 1980; Kramer et al., 1980) and are in the form of highly insoluble imorganic sulfides or sulfates (MacDonald, 1982), although chromium is associated initially with lignosulfonate. In a used drilling fluid more than 75 percent of the chrome lignosulfonate becomes bound to the clay fraction (Knox, 1978; McAtee and Smith, 1969; Skelly and Dieball, 1969). Heavy metals in the form of insoluble sulfides, adsorbed to particulates, or in the form of nonlabile organic complexes, have a much lower bioavailability to marine animals than do the metal ions in solution (Breteler et al., 1981; Bryan, 1982 Jenne and Luoma, 1977; Neff et al., 1978). Page et al. (1980) showed that mussels _Mytilus edulis_ accumulated more chromium from a solution of trivalent chromium salts than from solutions of ferrochrome lignosulfonate or aqueous fractions of chrome lignosulfonate drilling fluid. Capuzzo and Sasner (1977) showed that chromium adsorbed to bentonite clay was less bioavailable to mussels _Mytilus_ edulis and clams Mya arenaria than was an equivalent amount of chromium in a solution of $CrCl_3$. Chromium adsorbed to clay particles was much less available to sea scallops Placopecten magellanicus than chromium in solution (Liss et al., 1980). High levels of a metal in a sediment or drilling-fluid sample are not by themselves an indication of biological hazard. These adsorbed metals have very limited bioavailability.

Field studies conducted around offshore platforms report little to no significant elevation of metals in sediments (EG&G, 1982; Battelle/ Woods Hole Oceanographic Institution, 1983). This same pattern was seen around shallow and deep water, multiple well development and production platforms (8-25 wells per platform) in the Gulf of Mexico (Boothe and Presley, 1983). Boothe and Presley did note some slight elevation of mercury and lead within 125 m of two deep water locations. Based on analytical correlation, the mercury appeared to be associated with barium concentration and probably was due to trace contamination levels in the barite.

## Toxicity and Biomagnification

Several laboratory and field studies have addressed the uptake and retention by organisms of potentially toxic substances like trace metals and organic compounds in drilling fluids. The goal of these studies has been to determine whether marine organisms accumulate toxins in their tissues to concentrations sufficient to harm the organism or animals at higher trophic levels, including man. Laboratory studies have been useful in indicating uptake and depuration kinetics and, to a certain degree, the anatomical fates of accumulated materials, but laboratory studies of accumulation and field studies monitoring tissue are difficult to interpret because organisms may sequester and detoxify both metal and organic contaminants (Coombs and George, 1978; Jenkins and Brown, 1982; Stegeman, 1981). In order to effectively estimate the biological consequence of tissue or body burdens, it is important to examine the subcellular distributions of the contaminants (Bayne et al., 1980; Brown et al., 1982a; Jenkins et al., 1982). Because most bioaccumulation studies of drilling fluids have measured only total tissue or body burdens, their usefulness in predicting biological effects is limited. The little metal accumulation observed in both laboratory and field investigations, however, suggests that the biological effects of this accumulation are minimal.

Another issue that must be considered is the potential for the biomagnification of accumulated contaminant body burdens through marine food webs. This issue has not been addressed directly with regard to drilling fluids. Phelps et al. (1975) examined the distributions of heavy metals, however, particularly chromium, in Narragansett Bay organisms representing several trophic levels. Their data suggest that chromium body burdens decrease with trophic level. In similar studies in the Southern California Bight, Brown et al. (1982b) found zinc and copper levels decrease with higher trophic level, suggesting that inorganic metals do not biomagnify. In these studies, however, total mercury and DDT body burdens were found to increase significantly with

TABLE 23 Summary of Biological Effects of Drilling Fluids and Drilling Fluid Ingredients on Marine Animals

| | Laboratory Studies | |
|---|---|---|
| Parameter | Acute Lethal Bioassays (LC50 Range, ppm) | Chronic and Sublethal Effects[a] |
| Drilling fluid ingredients | | |
| Barite, bentonite, and lignite | >10,000 | 5-mm layer on sediment |
| Chrome- & Ferrochrome-lignosulfonates | 120-12,000 | 50 ppm |
| Chromium (VI) | 0.5-250 | 12 ppb |
| Diesel fuel | 0.1-1,000 | ≃10 ppb water, 100 ppm sediment |
| Paraformaldehyde | 0.07-30 | 10 ppb |
| Detergents, Surfactants | 0.4-14,000 | -- |
| Used drilling fluids | of 400 Bioassays<br>38% >100,000<br>41% 10,000-99,999<br>12% 1,000-9,999<br>6% 100-999<br>0.5% <100<br>3% LC50 Not Determinable | 1-160,000 ppm in water<br>1-12 mm Layer on bottom<br>50-100,000 ppm affects recruitment to microcosms; some bioaccumulation of barium and chromium demonstrated |
| | FIELD STUDIES | |
| Community responses | Effects seen only on benthos in the vicinity of discharge, and are most pronounced in low-energy environments where discharges accumulate on bottom. | |
| Bioaccumulation of metals[b] | Small uptake of barium and chromium immediately after drilling | |

[a]The lowest concentration at which effects are observed.
[b]There are no specific data available on the bioaccumulation of hydrocarbons.

increased trophic level. Of the total mercury measured, some 90 percent was organic (e.g., $CH_3Hg$). A more direct examination of the biomagnification of metals in the marine environment would be useful.

## CONCLUSIONS

Based on laboratory and field studies to date, most water-based drilling fluids used on the U.S. OCS have low acute and chronic toxicities to marine organisms in light of the fluids expected or observed rates of dilution and dispersal in the ocean after discharge. Their effects are restricted primarily to the ocean floor in the immediate vicinity and for a short distance downcurrent from the discharge. The bioaccumulation of metals from drilling fluids appears to be restricted to barium and chromium and is observed to be small in the field. Table 23 summarizes the concentrations of drilling fluids eliciting deleterious responses in marine organisms.

## REFERENCES

Abel, P.D. 1974. Toxicity of synthetic detergents to fish and aquatic invertebrates. J. Fish. Biol. 6:179-298.

American Petroleum Institute. 1978. Oil and gas well drilling fluid chemicals. API. Bull. 13F. 1st ed. American Petroleum Institute, Washington, D.C.

Anderson, J.W., R.G. Riley, and R.M. Bean. 1978. Recruitment of benthic animals as a function of petroleum hydrocarbon concentrations in the sediment. J. Fish. Res. Bd. Can. 35:776-790.

Atema, J., D.F. Leavitt, D.E. Barshaw, and M.C. Cuomo. 1982. Effects of drilling fluids on behavior of the American lobster, *Homarus americanus* in water column and substrate exposures. Can. J. Fish. Aquat. Sci. 39:675-690.

Augenfeld, J.M. 1980. Effects of Prudhoe Bay crude oil contamination on sediment working rates of *Abarenicola pacifica*. Mar. Environ. Res. 3:307-313.

Augenfeld, J.M., J.W. Anderson, R.G. Riley, and B.L. Thomas. 1982. The fate of polyaromatic hydrocarbons in an intertidal sediment exposure system: bioavailability to *Macoma inquinata* (Mollusca: Pelecypoda) and *Abarenicola pacifica* (Annelida: Polychaeta). Mar. Environ. Res. 7:31-50.

Augenfeld, J.M., J.W. Anderson, D.L. Woodruff and J.L. Webster. 1980. Effects of Prudhoe Bay crude oil-contaminated sediments on *Protothaca staminea* (Mollusca: Pelecypoda): hydrocarbon content, condition index, free amino acid level. Mar. Environ. Res. 4:135-143.

Ayers, R.C., Jr., T.C. Sauer, Jr., R.P. Meek, and G. Bowers. 1980a. An environmental study to assess the impact of drilling discharges in the Mid-Atlantic. I. Quantity and fate of discharges. In: Proceedings of a Symposium on Research on Environmental Fate and Effects of Drilling Fluids and Cuttings. Washington, D.C.: Courtesy Associates. Pp. 382-418.

Ayers, R.C., Jr., T.C. Sauer, Jr., D.O. Stuebner, and R.P. Meek. 1980b. An environmental study to assess the effect of drilling fluids on water quality parameters during high rate, high volume discharges to the ocean. In: Proceedings of a Symposium on Research on Environmental Fate and Effects of Drilling Fluids and Cuttings. Washington, D.C.: Courtesy Associates. Pp. 351-381.

Baker, J.M. (ed.). 1976. Marine Ecology and Oil Pollution. New York: Halsted Press. 566 pp.

Battelle/Woods Hole Oceanographic Institution. 1983. Georges Bank benthic infauna monitoring program. Final report, Year 1. Contract No. 14-12-0001-29192. Prepared for the New York OCS Office, Minerals Management Service, U.S. Department of the Interior.

Benech, S., R. Bowker, and B. Pimental. 1980. Chronic effects of drilling fluids exposure to fouling community composition on a semi-submersible exploratory drilling vessel. In: Proceedings of a Symposium on Research on Environmental Fate and Effects of Drilling Fluids and Cuttings. Washington, D.C.: Courtesy Associates. Pp. 611-635.

Birdsong, C.L., and L.W. Avault. 1971. Toxicity of certain chemicals to juvenile pompano. Prog. Fish-Cult. 33:76-80.

Boesch, D.F. 1979. Benthic ecological studies: macrobenthos. In: Middle Atlantic Outer Continental Shelf Environmental Studies. Vol. IIB. Chemical and Biological Benchmark Studies. Special Report in Applied Marine Science and Ocean Engineering No. 194. Virginia Institute of Marine Sciences, Gloucester Point, Va. New York OCS Office, Bureau of Land Management, U.S. Department of the Interior.

Boesch, D.F., and R. Rosenberg. 1981. Response to stress in marine benthic communities. In: G.W. Barrett and R. Rosenberg (eds.), Stress Effects on Natural Ecosystems. New York: John Wiley & Sons. Pp. 179-200.

Bookhout, C.G., R. Monroe, R. Forward, and J.D. Costlow, Jr. 1982. Effects of soluble fractions of drilling fluids and hexavalent chromium on the development of the crabs, *Rhithropanopeus harrisii* and *Callinectes sapidus*. EPA-600/S3-82-018. Final report to U.S. EPA Environmental Research Laboratory, Gulf Breeze, Fla.

Borthwick, P.W., and S.C. Schimmel. 1978. Toxicity of pentachlorophenol and related compounds to early life stages of selected estuarine animals. In: K.R. Rao (ed.), Pentachlorophenol: Chemistry, Pharmacology, and Environmental Toxicology. New York: Plenum Press. Pp. 141-146.

Bothner, M.H., R.R. Rendigs, E. Campbell, M.W. Doughton, P.J. Aruscavage, A.F. Dorrzapf, Jr., R.G. Johnson, C.M. Parmenter, M.J. Pikering, D.C. Brewster, and F.W. Brown. 1982. The Georges Bank monitoring program. Analysis of trace metals in bottom sediments. Interagency Agreement No. AA851-IA2-18. First year final report to the New York OCS Office, Minerals Management Service, U.S. Department of the Interior. Woods Hole, Mass: Geological Survey, U.S. Department of the Interior.

Brannon, A.C., and K.R. Rao. 1979. Barium, strontium and calcium levels in the exoskeleton, hepatopancreas and abdominal muscle of the grass shrimp *Palaemonetes pugio*: relation to molting and exposure to barite. Comp. Biochem. Physiol. 63A:261-274.

Breteler, R.J., I. Valiela, and J.M. Teal. 1981. Bioavailability of mercury in several northeastern U.S. *Spartina* ecosystems. Estuarine Coastal Shelf Sci. 12:155-166.

Brown, D.A., R.W. Gossett, and K.D. Jenkins. 1982a. Contaminants in white croakers *Geneyonemus lineatus* (Ayres, 1855) from the Southern California bight. II. Chlorinated hydrocarbon detoxification/toxification. In: W.B. Vernberg, A. Calabrese, F.P. Thurberg, and F.J. Vernberg (eds.). Physiological Mechanisms of Marine Pollution Toxicity. New York: Academic Press. In press.

Brown, D.A., R.W. Gossett, G.P. Hershelman, K.A. Schofer, K.D. Jenkins, and E.M. Perkins. 1982b. Bioaccumulation and detoxification of contaminants in marine organisms from southern California coastal waters. Proceedings of the Southern California Academy of Sciences. In press.

Bryan, G.W. In press. The biological availability and effects of heavy metals in marine deposits. In: Proceedings of Ocean Dumping Symposium. New York: Wiley-Interscience.

Cantelmo, F.R., M.E. Tagatz, and K.R. Rao. 1979. Effect of barite on meiofauna in a flow-through experimental system. Mar. Environ. Res. 2:301-309.

Capuzzo, J.M., and J.J. Sasner, Jr. 1977. The effect of chromium on filtration rates and metabolic activity of *Mytilus edulis* L. and *Mya arenaria* L. In: F.J. Vernberg, A. Calabrese, F.P. Thurberg, and W.B. Vernberg (eds.), Physiological Responses of Marine Biota to Pollutants. New York: Academic Press. Pp. 225-240.

Carls, M.G., and S.D. Rice. 1980. Toxicity of oil well drilling fluids to Alaskan larval shrimp and crabs. Research unit 72. Final report. Project No. R7120822. Outer Continental Shelf Energy Assessment Program. Bureau of Land Management, U.S. Department of the Interior. 29 pp.

Carney, L.L., and L. Harris. 1975. Thermal degradation of drilling mud additives. In: Proceedings, Environmental Aspects of Chemical Use in Well-Drilling Operations, May 21-23, 1975, Houston, Tex. Report No. EPA-5601/1-75-004. Washington, D.C.: Office of Toxic Substances, U.S. Environmental Protection Agency. vi + 604 pp.

Carr, R.S., W.L. McCulloch, and J.M. Neff. 1982. Bioavailability of chromium from a used chrome-lignosulfonate drilling fluids to five species of marine invertebrates. Mar. Environ. Res. 6:189-204.

Carr, R.S., L.A. Reitsema, and J.M. Neff. 1980. Influence of used chrome-lignosulfonate drilling fluids on the survival, respiration, growth, and feeding activity of the opossum shrimp *Mysidopsis almyra*. In: Proceedings of a Symposium on Research on Environmental Fate and Effects of Drilling Fluids and Cuttings. Washington, D.C.: Courtesy Associates. Pp. 944-963.

Chaffee, C., and R.B. Spies. 1982. The effects of used ferrochrome lignosulfonate drilling fluids from a Santa Barbara Channel oil well on the development of starfish embryos. Mar. Environ. Res. 7:265-277.

Chesser, B.G., and W.H. McKenzie. 1975. Use of a bioassay test in evaluating the toxicity of drilling fluid additives on Galveston Bay Shrimp. In: Proceedings, Environmental Aspects of Chemical Use in Well-Drilling Operations, May 21-23, 1975, Houston, Tex. Report No. EPA-560/1-75-004. Washington, D.C.: Office of Toxic Substances, U.S. Environmental Protection Agency. Pp. 153-168.

Choi, D.R. 1982. Coelobites (reef cavity-dwellers) as indicators of environmental effects caused by offshore drilling. Bull. Mar. Sci. 32:880-889.

Chow, T.S., and C.B. Snyder. 1980. Barium in the marine environment. A potential indicator of drilling contamination. In: Proceedings of a Symposium on Research on Environmental Fate and Effects of Drilling Fluids and Cuttings. Washington, D.C.: Courtesy Associates. Pp. 723-738.

Conklin, P.J., D.G. Doughtie, and K.R. Rao. 1980. Effects of barite and used drilling fluids on crustaceans, with particular reference to the grass shrimp, *Palaemonetes pugio*. In: Proceedings of a Symposium on Research on Environmental Fate and Effects of Drilling Fluids and Cuttings. Washington, D.C.: Courtesy Associates. Pp. 723-738.

Conklin, P.J., D. Drysdale, D.G. Doughties, K.R. Rao, J.P. Kakareka, T.R. Gilbert, and R.F. Shokes. In press. Comparative toxicity of drilling fluids: role of chromium and petroleum hydrocarbons. Mar. Environ. Res.

Coombs, T.L., and S.C. George. 1978. Mechanisms of immobilization and detoxification of metals in marine organisms. In: D.S. McLusky and A.J. Berry (eds.). Physiology and Behavior of Marine Organisms. Oxford: Pergamon Press. Pp. 179-187.

Craddock, D.R. 1977. Acute toxic effects of petroleum on arctic and subarctic marine organisms. In: D.C. Malins (eds.), Effects of Petroleum on Arctic and Subarctic Marine Environments and Organisms. Vol. II. Biological Effects. New York: Academic Press. Pp. 1-93.

Cranston, R.E., and J.W. Murray. 1980. Chromium species in the Columbia River and estuary. Limnol. Oceanogr. 25:1104-1112.

Crawford, R.B., and J.D. Gates. 1981. Effects of drilling fluid on the development of a teleost and an echinoderm. Bull. Environ. Contam. Toxicol. 26:207-212.

Crippen, R.W., S.L. Hodd, and G. Greene. 1980. Metal levels in sediment and benthos resulting from a drilling fluid discharge into the Beaufort Sea. In: Proceedings of a Symposium on Research on Environmental Fate and Effects of Drilling Fluids and Cuttings. Washington, D.C.: Courtesy Associates. Pp. 636-669.

Dames & Moore, Inc. 1978. Drilling fluid dispersion and biological effects study for the lower Cook Inlet C.O.S.T. well. Report submitted to Atlantic Richfield Co. Dames & Moore, Inc., Anchorage, Alaska. 309 pp.

Daugherty, F.W. 1951. Effects of some chemicals used in oil well drilling on marine animals. Sewage Ind. Wastes 23:1282-1287.

Derby, C.D., and J. Atema. 1981. Influence of drilling fluids on the primary chemosensory neurons in walking legs of the lobster, _Homarus americanus_. Can. J. Fish. Aquat. Sci. 38:268-274.

Dodge, R.E. 1982. Effects of drilling fluids on the reef-building coral _Montastrea annularis_. Mar. Biol. 71:141-147.

Ecomar, Inc. 1978. Tanner Bank fluids and cuttings study. Conducted for Shell Oil Company, January through March 1977. Ecomar, Inc., Goleta, Calif. 495 pp.

EG&G Bionomics. 1976a. Acute toxicity of two-drilling fluid components, barite and Aquagel, to the marine alga (_Skeletonema_ costatum) and calanoid copepods (Acartia tonsa). Report submitted to Shell Oil Co., New Orleans. EG&G Bionomics, Pensacola, Fla.

EG&G Bionomics. 1976b. Toxicity of ICMO Services No. 1 drilling fluids to a marine algae (Skeletonema costatum) and calanoid copepod (Acartia tonsa). Toxicity report to IMCO Services. EG&G Marine Research Laboratory, Pensacola, Fla. 16 pp.

EG&G Bionomics. 1976c. Acute toxicity test report submitted to Shell Oil Co., New Orleans, La.

EG&G Environmental Consultants. 1980. Monitoring program for Exxon's block 564 Jacksonville OCS area (Lease OCS-G3705). Report submitted to Exxon Company, U.S.A., Houston, Tex.

EG&G Environmental Consultants. 1982. A study of environmental effects of exploratory drilling on the mid-Atlantic outer continental shelf--final report of the Block 684 Monitoring Program. Waltham, Mass.: Available from Offshore Operators Committee, Environmental Subcommittee, P.O. Box 50751, New Orleans, LA 70150.

Eisler, R., and R.J. Hennekey. 1977. Acute toxicitiies of $Cd^{2+}$, $Cr^{6+}$, $Hg^{2+}$, $Ni^{2+}$, and $Zn^{2+}$ to estuarine macrofauna. Arch. Environ. Contam. Toxicol. 6:315-323.

EPA/COE. 1977. Ecological evaluation of proposed discharge of dredged material into ocean waters. EPA/COE Technical Committee on Criteria for Dredged and Fill Material. U.S. Army Corps of Engineers, Waterways Experiment Station, Vicksburg, Miss.

ERCO. 1980. Results of joint bioassay monitoring program. Final report to the Offshore Operators Committee under direction of Exxon Production Research Co., Houston, Tex. ERCO, Inc., Cambridge, Mass.

Espey, Huston & Associates, Inc. 1981. Bioassay and depuration studies on two types of barite. Document No. 81123. Report to Magcobar Group, Dresser Industries, Inc., Houston, Tex. Espey, Huston & Assoc., Inc., Houston, Tex. 256 pp.

Federal Register. 1978. 43(49). Monday, 13 March. Pp. 10474-10508.

Frank, P.M., and P.B. Robertson. 1979. The influence of salinity on toxicity of cadmium and chromium to the blue crab, Callinectes sapidus. Bull. Environ. Contam. Toxicol. 21:74-78.

George, S.G., B.J.S. Pirie, and T.L. Coombs. 1976. The kinetics of accumulation and excretion of ferric hydroxide in Mytilus edulis (L.) and its distribution in the tissues. J. Exp. Mar. Biol. Ecol. 23:71-84.

Gerber, R.P., E.S. Gilfillan, J.R. Hotham, L.J. Galletto, and S.A. Hanson. 1981. Further studies on the short and long-term effect of used drilling fluids on marine organisms. Unpublished. Final Report, Year II to the American Petroleum Institute, Washington, D.C. 30 pp.

Gerber, R.P., E.S. Gilfillan, B.T. Page, D.S. Page, and J.B. Hotham. 1980. Short- and long-term effects of used drilling fluids on marine organisms. In: Proceedings of a Symposium on Research on Environmental Fate and Effects of Drilling Fluids and Cuttings. Washington, D.C.: Courtesy Associates. Pp. 882-911.

Gettleson, D.A. 1978. Ecological impact of exploratory drilling: a case study. In: Energy/Environment '78. Society of Petroleum Industry Biologists Symposium, 22-24 August, 1978, Los Angeles, Calif. 23 pp.

Gettleson, D.A., and C.E. Laird. 1980. Benthic barium in the vicinity of six drill sites in the Gulf of Mexico. In: Proceedings of a Symposium on Research on Environmental Fate and Effects of Drilling Fluids and Cuttings. Washington, D.C.: Courtesy Associates. Pp. 739-788.

Gilbert, T.R. 1981. A study of the impact of discharged drilling fluids on the Georges Bank environment. New England Aquarium. H. E. Edgerton Research Laboratory. Progress Rept. No. 2 to U.S. EPA, Gulf Breeze, Fla. 98 pp.

Gillmor, R.B., C.A. Menzie, G.M. Mariani, D.R. Levin, R.C. Ayers, Jr., and T.C. Sauer, Jr. In press. Effects of exploratory drilling discharges on the benthic environment in the middle Atlantic OCS: biological results of a one-year postdrilling survey. In: Proceedings of Ocean Dumping Symposium. New York: Wiley-Interscience.

Gillmor, R.B., C.A. Menzie, and J. Ryther, Jr. 1981. Side-scan sonar and TV observations of the benthic environment and megabenthos in the vicinity of an OCS exploratory well in the Middle Atlantic Bight. In: Conference Record. Vol. 2 Oceans '81 Symposium. Boston, Mass. IEEE Publ. No. 81CH1685-7 IEEE Service Center, Piscataway, N.J. Pp. 727-731.

Grantham, C.K., and J.P. Sloan. 1975. Toxicity study of drilling fluid chemicals on aquatic life. In: Environmental Aspects of Chemical Use in Well-Drilling Operations. EPA/560/1-75-004. Washington, D.C.: U.S. Environmental Protection Agency. Pp. 103-112.

Grassle, J.F., R.E. Imgren, and J.P. Grassle. 1980. Response of benthic communities in MERL experimental ecosystems to low level, chronic additions of No. 2 fuel oil. Mar. Environ. Res. 4:279-297.

Hollingsworth, J.W., and R.A. Lockhart. 1975. Fish toxicity of dispersed clay drilling mud deflocculants. In: Environmental Aspects of Chemical Use in Well-Drilling Operations. EPA-360/1-75-004. Washington, D.C.: U.S. Environmental Protection Agency. Pp. 113-123.

Houghton, J.P., D.L. Beyer, and E.D. Thielk. 1980. Effects of oil well drilling fluids on several important Alaskan marine organisms. In: Proceedings of a Symposium on Research on Environmental Fate and Effects of Drilling Fluids and Cuttings. Washington, D.C.: Courtesy Associates. Pp. 1017-1043.

Hudson, J.H., and D.M. Robbin. 1980. Effects of drilling fluids on the growth rate of the reef-building coral, Montastrea annularis. In: Proceedings of a Symposium on Research on Environmental Fate and Effects of Drilling Fluids and Cuttings. Washington, D.C.: Courtesy Associates. Pp. 1101-1122.

Hudson, J.H., E.A. Shinn, and D.M. Robbin. 1982. Effects of offshore oil drilling on Philippine Reef corals. Bull. Mar. Sci. 32:890-908.

Intergovernmental Maritime Consultative Organization, et al. 1969. Abstract of first session report of Joint Group of Experts on the Scientific Aspects of Marine Pollution. Water Res. 3:995-1005.

Jenkins, K.D., and D.A. Brown. 1982. Determining the biological significance of contaminant bioaccumulation. In: H. White (ed.), Proceedings of the Workshop on Meaningful Measures of Marine Pollution Effects. Pensacola, Fla., April 1982. College Park, Md.: University of Maryland Sea Grant Program. In press.

Jenkins, K.D., D.A. Brown, G.P. Hershelman, and W.C. Meyer. 1982. Contaminants in white croakers Geneyonemus lineatus (Avres, 1855) from the Southern California bight. I. Trace metals detoxification/toxification. In: W.B. Vernberg, A. Calabrese, F.P. Thurberg, and F.J. Vernberg (eds.). Physiological Mechanisms of Marine Pollutant Toxicity. New York: Academic Press. In press.

Jenne, E.A., and S.N. Luoma. 1977. Forms of trace elements in soils, sediments, and associated waters: an overview of their determination and biological availability. In: H. Drucker and R.E. Wildung (eds.), Biological Implications of Metals in the Environment. NTIS Conf.-750929. Springfield, Va. Pp. 110-143.

Jones, M., and M. Hulse. 1982. Drilling fluid bioassays and the OCS. Oil Gas J. 80(25):241-244.

Knox, F. 1978. The behavior of ferrochrome lignosulfonate in natural waters. Masters Thesis. Massachusetts Institute of Technology, Cambridge, Mass. 65 pp.

Kramer, J.R., H.D. Grundy, and L.G. Hammer. 1980. Occurrence and solubility of trace metals in barite for ocean drilling operations. In: Proceedings of a Symposium on Research on Environmental Fate and Effects of Drilling Fluids and Cuttings. Washington, D.C.: Courtesy Associates. Pp. 789-798.

Krone, M.A., and D.C. Biggs. 1980. Sublethal metabolic responses of the hermatypic coral Madracis decactis exposed to drilling fluids, enriched with ferrochrome lignosulfonate. In: Proceedings of a Symposium on Research on Environmental Fate and Effects of Drilling Fluids and Cuttings. Washington, D.C.: Courtesy Associates. Pp. 1079-1100.

Land, B. 1974. The toxicity of drilling fluid components to aquatic biological systems. A literature review. Tech. Rep. No. 487. Environment Canada, Fisheries and Marine Service, Research and Development Directorate, Freshwater Institute, Winnipeg, Manitoba, Canada. 33 pp.

Lees, D.C., and J.P. Houghton. 1980. Effects of drilling fluids on benthic communities at the lower Cook Inlet C.O.S T. well. In: Proceedings of a Symposium on Research on Environmental Fate and Effects of Drilling Fluids and Cuttings. Washington, D.C.: Courtesy Associates. Pp. 209-350.

Liss, R.G., F. Knox, D. Wayne, and T.R. Gilbert. 1980. Availability of trace elements in drilling fluids to the marine environment. In: Proceedings of a Symposium on Research on Environmental Fate and Effects of Drilling Fluids and Cuttings. Washington, D.C.: Courtesy Associates. Pp. 691-722.

Logan, W.J., J.B. Sprague, and B.D. Hicks. 1973. Acute lethal toxicity to trout of drilling fluids and their constituent chemicals as used in the Northwest Territories. Appendix to M.R. Falk and M.J. Lawrence. Acute toxicity of petrochemical drilling fluids components and wastes to fish. Tech. Rep. Ser. No. CEN-T-73-1. Environment Canada, Resource Management Branch, Central Region, Winnipeg, Manitoba, Canada. Pp. 45-108.

Lyes, M.C. 1979. Bioavailability of hydrocarbon from water and sediment to the marine worm Arenicola marina. Mar. Biol. 55:121-127.

MacDonald, R.W. 1982. An examination of metal inputs to the southern Beaufort Sea by disposal of waste barite in drilling fluid. Ocean Manage. 8:29-49.

Malins, D.C. (ed.). 1977. Effects of Petroleum on Arctic and Subarctic Marine Environments and Organisms. Vol. II. Biological Effects. New York: Academic Press. 500 pp.

Mariani, G.M., L.V. Sick, and C.C. Johnson. 1980. An environmental monitoring study to assess the impact of drilling dischages in the mid-Atlantic. III. Chemical and physical alterations in the benthic environment. In: Proceedings of a Symposium on Environmental Fate and Effects of Drilling Fluids and Cuttings. Washington, D.C.: Courtesy Associates. Pp. 438-498.

Marine Bioassay Labs. 1982. Drilling fluids bioassays. Texaco Habitat Platform Well A-1 Pitas Point Lease Sale Unit: Acanthomysis sculpta and Macoma nasuta. Report submitted to Texaco, Inc., Los Angeles, Calif., and IMCO Services, Houston, Tex., by Marine Bioassay Labs., Watsonville, Calif.

Maurer, D., W. Leathem, and C. Menzie. 1981. The impact of drilling fluid and well cuttings on polychaete feeding guilds from the U.S. northeastern continental shelf. Mar. Pollut. Bull. 12:342-347.

McAtee, J.L., and N.R. Smith. 1969. Ferrochrome lignosulfonates. I. X-ray absorption edge fine structure spectroscopy. II. Interaction with ion exchange resin and clays. J. Colloid Interface Sci. 29:389-398.

McCain, B.B., H.O. Hodgins, W.D. Gronlund, J.W. Hawkes, D.W. Brown, M.S. Myers, and J.J. Vandermuelen. 1978. Bioavailability of crude oil from experimentally oiled sediments to English sole (Parophrys vetulus), and pathological consequences. J. Fish. Res. Bd. Can. 35:657-664.

McCulloch, W.L., J.M. Neff, and R.S. Carr. 1980. Bioavailability of heavy metals from used offshore drilling fluids to the clam Rangia cuneata and the oyster Crassostrea gigas. In: Proceedings of a Symposium on Research on Environmental Fate and Effects of Drilling Fluids and Cuttings. Washington, D.C.: Courtesy Associates. Pp. 964-983.

McFarland, V.A., and R.K. Peddicord. 1980. Lethality of a suspended clay to a diverse selection of marine and estuarine macrofauna. Arch. Environ. Contam. Toxicol. 9:733-741.

McLeay, D.J. 1976. Marine toxicity studies on drilling fluid wastes. In: Industry/Government Working Group in Disposal Waste Fluids from Petroleum Exploratory Drilling in the Canadian Arctic. Vol. 10. Yellowknife, N.W.T., Canada. 17 pp.

Meek, R.P., and J.P. Ray. 1980. Induced sedimentation, accumulation, and transport resulting from exploratory drilling discharges of drilling fluids and cuttings. In: Proceedings of a Symposium on Research on Environmental Fate and Effects of Drilling Fluids and Cuttings. Washington, D.C.: Courtesy Associates. Pp. 259-284.

Menzie, C.A., D. Maurer, and W.A. Leathem. 1980. An environmental monitoring study to assess the impact of drilling discharges in the mid-Atlantic. IV. The effects of drilling discharges on the benthic community. In: Proceedings of a Symposium on Environmental Fate and Effects of Drilling Fluids and Cuttings. Washington, D.C.: Courtesy Associates. Pp. 499-540.

Moraitou-Apostolopoulou, M., and G. Verriopoulos. 1982. Toxicity of chromium to the marine planktonic copepod *Acartia clausi*, Giesbrecht. Hydrobiologia 96:121-127.

Morse, M.P., W.E. Robinson, and W.E. Wehling. 1982. Effects of sublethal concentrations of the drilling fluids components attapulgite and Q-Broxin on the structure and function of the gill of the scallop, *Placopecten magellanicus* (Gmelin). In: W.B. Vernberg, A. Calabrese, F.P. Thurberg and F.J. Vernberg, (eds.). Physiological Mechanisms of Marine Pollutant Toxicity. New York: Academic Press. Pp. 235-260.

Nakayama, E., T. Kuwamoto, S. Tsurubo, H. Tokoro, and T. Fujinaga. 1981a. Chemical speciation of chromium in seawater. Part 1. Effect of naturally occurring organic materials on the complex formation of chromium (III). Analy. Chim. Acta 130:289-294.

Nakayama, E., T. Kuwamoto, S. Tsurubo, and T. Fujinaga. 1981b. Chemical speciation of chromium in seawater. Part 2. Effects of manganese oxides and reducible organic materials on the redox processes of chromium. Analy. Chim. Acta 130:401-404.

Nakayama, E., T. Kuwamoto, H. Tokoro, and T. Fujinaga. 1981c. Chemical speciation of chromium in seawater. Part 3. The determination of chromium species. Analy. Chim. Acta 131:247-254.

National Research Council. 1975. Petroleum in the Marine Environment. Washington, D.C.: National Academy of Sciences.

Neff, J.M. 1979. Polycyclic Aromatic Hydrocarbons in the Aquatic Environment. Sources, Fates, and Biological Effects. Barking Essex, England: Applied Science Publ. 262 pp.

Neff, J.M. 1980. Effects of used drilling fluids on benthic marine animals. Publ. No. 4330. American Petroleum Institute, Washington, D.C. 31 pp.

Neff, J.M., and J.W. Anderson. 1981. Response of Marine Animals to Petroleum and Specific Petroleum Hydrocarbons. New York: Halsted Press. 177 pp.

Neff, J.M., R.S. Carr, and W.L. McCulloch. 1981. Acute toxicity of a used chrome lignosulfonate drilling fluids to several species of marine invertebrate. Mar. Environ. Res. 4:251-266.

Neff, J.M., R.S. Foster, and J.F. Slowey. 1978. Availability of sediment-adsorbed heavy metals to benthos with particular emphasis on deposit-feeding infauna. Tech. Rep. D-78-42. U.S. Army Engineer Waterways Experiment Station, Vicksburg, Miss. 286 pp.

Neff, J.M., W.L. McCulloch, R.S. Carr, and K.A. Retzer. 1980. Comparative toxicity of four used offshore drilling fluids to several species of marine animals from the Gulf of Mexico. In: Proceedings of a Symposium on Research on Fate and Effects of Drilling Fluids and Cuttings. Washington, D.C.: Courtesy Associates. Pp. 866-881.

Nelson, D.W., S. Liu, and L.E. Sommers. 1980. Plant uptake of toxic metals present in drilling fluids. In: Proceedings of a Symposium on Research on Fate and Effects of Drilling Fluids and Cuttings. Washington, D.C.: Courtesy Associates. Pp. 114-138.

Nimmo, D. et al. 1979. The long term effects of suspended particulates on survival and reproduction of the mysid shrimp, Mysidopsis bahia, in the laboratory. In: Proceedings of MESA Symposium, New York, June 10-15. Rockville, Md.: National Oceanic and Atmospheric Administration.

Norseth, T. 1981. The carcinogenicity of chromium. Environ. Health Perspect. 40:121-130.

Northern Technical Services. 1981. Beaufort Sea drilling effluent disposal study. Performed for the Reindeer Island stratigraphic test well participants under the direction of Sohio Alaska Petroleum Company. Northern Technical Services, Anchorage, Alaska. Available from Sohio Alaska Petroleum Co., Anchorage, Alaska. 329 pp.

Oshida, P.S., L.S. Word, and A.J. Mearns. 1981. Effects of hexavalent and trivalent chromium on the reproduction of Neanthes arenaceodentata (Polychaeta). Mar. Environ. Res. 5:41-50.

Oviatt, C., J. Frithsen, J. Gearing, and P. Gearing. 1982. Low chronic additions of No. 2 fuel oil: chemical behavior, biological impact and recovery in a simulated estuarine environment. Mar. Ecol. Prog. Ser. 9:121-136.

Paffenhofer, R. 1972. The effects of suspended "red mud" on mortality, body weight, and growth of the marine planktonic copepod, Calanus helgolandicus. Water Air Soil Pollut. 1:314-321.

Page, D.S., B. T. Page, J.R. Hotham, E.S. Gilfillan, and R.P. Gerber. 1980. Bioavailability of toxic constituents of used drilling fluids. In: Proceedings of a Symposium on Research on Environmental Fate and Effects of Drilling Fluids and Cuttings. Washington, D.C.: Courtesy Associates. Pp. 984-996.

Payne, J.R., J.L. Lambach, R.E. Jordan, G.D. McNabb, Jr., R.R. Sims, A. Abasumara, J.G. Sutton, D. Generro, S. Gagner, and R.F. Shokes. 1982. Georges Bank monitoring program. Analysis of hydrocarbons in bottom sediments and analysis of hydrocarbons and trace metals in benthic fauna. First year final report to the New York OCS Office, Minerals Management Service, U.S. Department of the Interior, by Science Applications, Inc., La Jolla, Calif.

Pearson, R.G. 1981. Recovery and recolonization of coral reefs. Mar. Ecol. Prog. Ser. 4:105-122.

Perricone, D. 1980. Major drilling fluid additives. In: Proceedings of a Symposium on Research on Environmental Fate and Effects of Drilling Fluids and Cuttings. Washington, D.C.: Courtesy Associates. Pp. 15-29.

Petrazzuolo, G. 1981. Preliminary report. An environmental assessment of drilling fluids and cuttings released onto the Outer Continental Shelf. Vol. 1: Technical assessment. Vol. 2: Tables, figures and Appendix A. Draft report prepared for Industrial Permits Branch, Office of Water Enforcement and Ocean Programs Branch, Office of Water and Waste Management, U.S. Environmental Protection Agency, Washington, D.C.

Phelps, D.K., G. Telek, and R.L. Lapan. 1975. Assessment of heavy metal distribution within the food web. In: Pearson and Frangipane (eds.). Marine Pollution and Marine Waste Disposal. New York: Pergamon Press.

Powell, E.N., M. Kasschau, E. Che, M. Loenig, and J. Peron. 1982. Changes in the free amino acid pool during environmental stress in the gill tissue of the oyster, Crassostrea virginica. Comp. Biochem. Physiol. 71A:591-598.

Reish, D.J., J.M. Martin, F.M. Piltz, and J.L. Ward. 1976. The effect of heavy metals on laboratory populations of two polychaetes with comparisons to the water quality conditions and standards in Southern California marine waters. Water. Res. 10:299-302.

Rhoads, D.C., P.L. McCall, and J.Y. Yingst. 1981. Disturbance and production of the estuarine seafloor. Am. Sci. 66:577-586.

Rice, S.D., J.W. Short, and J.F. Karinen. 1977. Comparative oil toxicity and comparative animal sensitivity. In: D.A. Wolfe (ed.), Fate and Effects of Petroleum Hydrocabons in Marine Organisms and Ecosystems. New York: Pergamon Press. Pp. 78-94.

Roesijadi, G., and J.W. Anderson. 1979. Condition index and free amino acid content of Macoma inquinata exposed to oil-contaminated marine sediments. In: W.B. Vernberg, F.P. Thurberg, A. Calabrese and F.J. Vernberg (eds.), Marine Pollution: Functional Responses. New York: Academic Press. Pp. 69-83.

Roesijadi, G., J.W. Anderson, and J.W. Blaylock. 1978a. Uptake of hydrocarbons from marine sediments contaminated with Prudhoe Bay crude oil: influence of feeding type of test species and availability of polycyclic aromatic hydrocarbons. J. Fish. Res. Bd. Can. 35:608-614.

Roesijadi, G., D.L. Woodruff, and J.W. Anderson. 1978b. Bioavailability of naphthalenes from marine sediments artificially contaminated with Prudhoe Bay crude oil. Environ. Pollut. 15:223-229.

Rossi, S.S. 1977. Bioavailability of petroleum hydrocarbons from water, sediments and detritus to the marine annelid Neanthes arenaceodentata. 1977. In: Proceedings 1977 Oil Spill Conference (Prevention, Behavior, Control, Cleanup). Washington, D.C.: American Petroleum Institute. Pp. 621-626.

Rubinstein, N.I., R. Rigby, and C.N. D'Asaro. 1980. Acute and sublethal effects of whole used drilling fluids on representative estuarine organisms. In: Proceedings of a Symposium on Research on Environmental Fate and Effects of Drilling Fluids and Cuttings. Washington, D.C.: Courtesy Associates. Pp. 828-846.

Sanders, H.L., J.F. Grassle, G.R. Hampson, L.S. Morse, S. Garner-Price, and C.C. Jones. 1980. Anatomy of an oil spill: long-term effects from the grounding of the barge Florida off west Falmouth, Massachusetts. J. Mar. Res. 38:265-380.

Sharp, J.R., R.S. Carr, and J.M. Neff. 1982. Influence of used chrome lignosulfonate drilling and fluids on the early life history of the mummichog Fundulus heteroclitus. In: Proceedings of Ocean Dumping Symposium. New York: John Wiley & Sons.

Siferd, C.A. 1976. Drilling fluids wastes characteristics from drilling operations in the Canadian North. In: Industry/Government Working Group "A" Program, Pollution Aspects from Waste Drilling Fluids in the Canadian North. Vol. 5. Arctic Petroleum Operators Association and Environment Canada, Yellowknife, N.W.T., Canada.

Skelly, W.G., and D.E. Dieball. 1969. Behavior of chromate in drilling fluids containing chromate. In: Proceedings of the 44th Annual Meeting of the Society of Petroleum Engineers of AIME. Paper No. SPE 2539. 6 pp.

Smillie, R.H., K. Hunter, and M. Loutit. 1981. Reduction of chromium (VI) by bacterially produced hydrogen sulfide in a marine environment. Water Res. 15:1351-1354.

Smith, G.A., J.S. Nickels, R.J. Bobbie, N.L. Richards, and D.C. White. 1982. Effects of oil and gas well-drilling fluids on the biomass and community structure of microbiota that colonize sands in running seawater. Arch. Environ. Contam. Toxicol. 11:17-24.

Sprague, J.B., and W.J. Logan. 1979. Separate and joint toxicity to rainbow trout of substances used in drilling fluids for oil exploration. Environ. Pollut. 19:269-281.

Stegeman, J.J. 1981. Polynuclear aromatic hydrocarbons and their metabolism in the marine environment. In: H.V. Gelboin and P.O.P. Ts'o (eds.), Polycyclic Hydrocarbons and Cancer. Vol. 3. New York: Academic Press. Pp. 1-60.

Szmant-Froelich, A., V. Johnson, T. Hoen, J. Battey, G.J. Smith, E. Fleishmann, J. Porter, and D. Dallmeyer. In press. The physiological effects of oil-drilling fluids on the Carribbean coral _Montastrea annularis_. In: Proceedings of the 4th International Coral Reef Symposium, Manila. Vol. 1.

Tagatz, M.E., J.M. Ivey, C.E. DalBo, and J.L. Oglesby. 1982. Responses of developing macrobenthic communities to drilling fluids. Estuaries 5:131-137.

Tagatz, M.E., J.M. Ivey, H.K. Lehman, and J.L. Oglesby. 1978. Effects of lignosulfonate-type drilling fluids on development of experimental estuarine macrobenthic communities. Northeast. Gulf Sci. 2:25-42.

Tagatz, M.E., J.M. Ivey, H.K. Lehman, M. Tobia, and J.L. Oglesby. 1980. Effects of drilling fluids on development of experimental estuarine macrobenthic communities. In: Proceedings of a Symposium on Research on Environmental Fate and Effects of Drilling Fluids and Cuttings. Washington, D.C.: Courtesy Associates. Pp. 847-865.

Tagatz, M.E., J.M. Ivey, and J.L. Oglesby. 1979. Toxicity of drilling fluids biocides to developing macrobenthic communites. Northeast. Gulf Sci. 3:88-95.

Tagatz, M.E., and M. Tobia. 1978. Effect of barite ($BaSO_4$) on development of estuarine communites. Estuarine Coastal Mar. Sci. 7:401-407.

Thompson, J.H., Jr. 1978. Effects of drilling fluids on seven species of reef building corals as measured in field and laboratory. Final report to U.S.G.S. Grant No. 14-08-001-1627. College of Geosciences, Texas A&M University. 54 pp.

Thompson, J.H., Jr. 1980. Responses of selected scleractinian corals to drilling fluid used in the marine environment. Ph.D. Dissertation. Texas A&M University. 130 pp.

Thompson, J.H., Jr., and T.J. Bright. 1977. Effects of drilling fluids on sediment clearing rates of certain hermatypic corals. In: Oil Spill Conference. Washington, D.C.: American Petroleum Institute. Pp. 495-498.

Thompson J.H., Jr., and T.J. Bright. 1980. Effects of an offshore drilling fluid on selected corals. In: Proceedings of a Symposium on Research on Environmental Fate and Effects of Drilling Fluids and Cuttings. Washington, D.C.: Courtesy Associates. Pp. 1044-1078.

Thompson, J.H., Jr., E.A. Shinn, and T.J. Bright. 1980. Effects of drilling fluids on seven species of reef-building corals as measured in the field and laboratory. In: R.A. Geyer (ed.), Marine Environmental Pollution. Vol. 1. Hydrocarbons. Elsevier Oceanography Series. Vol. 27A. New York: Elsevier/North-Holland.

Thorson, G. 1957. Bottom communities (sublittoral or shallow shelf). In: J.W. Hedgepeth (ed.), Treatise on Marine Ecology and Paleoecology. Vol. I. Geol. Soc. Am. Mem. 67. Washington, D.C.

Thorson, G. 1966. Some factors influencing the recruitment and establishment of marine benthic communities. Neth. J. Sea. Res. 3:267-293.

Tillery, J.B., and R.E. Thomas. 1980. Heavy metal contamination from petroleum production platforms in the Gulf of Mexico. In: Symposium on Environental Fate and Effects of Drilling Fluids and Cuttings. Washington, D.C.: Courtesy Associates. Pp. 562-587.

Tornberg, L.D., E.D. Thielk, R.E. Nakatoni, R.C. Miller and S.O. Hillman. 1980. Toxicity of drilling fluids to marine organisms in the Beaufort Sea, Alaska. In: Proceedings of a Symposium on Research on Environmental Fate and Effects of Drilling Fluids and Cuttings. Washington, D.C.: Courtesy Associates. Pp. 997-1006.

Trefry, J.H., R.P. Trocine, and D.B. Meyer. 1981. Tracing the fate of petroleum drilling fluids in the northwest Gulf of Mexico. In: Oceans '81. Available from Marine Technology Society, Washington, D.C. Pp. 732-736.

Trocine, R.P., J.H. Trefry, and D.B. Meyer. 1981. Inorganic tracers of petroleum drilling fluid dispersion in the northwest Gulf of Mexico. Reprint Extended Abstract. Div. Environ. Chem. ACS Meeting. Atlanta, Ga. March-April 1981.

Van der Weijden, C.H., and M. Reith. 1982. Chromium(III)-Chromium(VI) interconversions in seawater. Mar. Chem. 11:565-572.

Ward, J.A. 1977. Chemoreception of heavy metals by the polychaetous annelid Myxicola infundibulum (Sabellidae). Comp. Biochem. Physiol. 58C:103-106.

Wheeler, R.B., J.B. Anderson, R.R. Schwrzer, and C.L. Hokanson. 1980. Sedimentary processes and trace metal contaminants in the Buccaneer oil/gas field, northwesten Gulf of Mexico. Environ. Geol. 3:163-175.

White, D.C., J.S. Nickels, M.J. Gehron, J.H. Parker, R.F. Martz, and N.L. Richards. In press. Effect of well-drilling fluids on the physiological status and microbial infection of the reef-building coral _Monastrea annularis_. Arch. Environ Contam. Toxicol.

Wilbert, T. 1971. Turbidity. In: Marine Ecology. New York: Wiley-Interscience. Pp. 1184-1194.

Wildish, D.J. 1972. Acute toxicity of polyoxyethylene esters and polyoxyethylene ethers to _S. salar_ and _G. oceanicus_. Water Res. 6:759-762.

Zitko, V. 1975. Toxicity and environmental properties of chemicals used in well-drilling operations. In: Environmental Aspects of Chemical Use in Well-Drilling Operations. EPA-560/1-75-004. Washington, D.C.: U.S. Environmental Protection Agency. Pp. 311-326.

# 5
# Considerations in Using the Information Available on the Fates and Effects of Drilling Discharges

Most information about the effects of drilling fluids on marine organisms comes from laboratory experiments. Most of these have studied lethal effects over a short period of time, typically in 96 hr LC50 toxicity tests. The organisms most frequently used in these bioassay tests have been the coastal and estuarine species readily available for testing and easily maintained in the laboratory. Only a few assessments of drilling-fluid effects have been made in the field, and these field measurements are not very advanced.

The limitations of the laboratory experiments have led to some criticisms of their adequacy and of their applicability in assessing the effects of drilling-fluid discharges on the OCS. These criticisms also apply to current assessments of the effects of most anthropogenic additions to the marine environment. Information on the effects of discharged drilling fluids is generally no less substantial than that on municipal and industrial wastes, sewage sludge, and dredged sediments and in some respects is of higher quality because of more sophisticated research in recent years. The issue to address is what degree of confidence is warranted by hazard assessment models that rely on laboratory studies of toxic effects along with predicted exposure regimes for the benthic and pelagic communities of the various continental shelf environments. In such models, testing acute toxicity is only the first step in evaluating biological effects. More sophisticated measures of environmental effects, some of which are discussed below, are required in rigorous models. Even in sophisticated investigations, however, a fundamental dilemma remains in relying on either prospective studies, which may be limited in their environmental realism, or retrospective field asssessments, which may be limited in their predictive value.

## LABORATORY EVALUATIONS OF TOXICITY

In evaluating the toxic effects of substances on aquatic organisms, two types of tests are used: (1) acute toxicity tests, which determine the concentration that causes the mortality of some proportion of test organisms, (for example, half in the LC50 test); and (2) chronic toxicity tests, which determine what concentration causes some other measurable effect. Acute toxicity tests are usually conducted during a 4-day period (96 h) to provide a standard for comparing the toxicities of different substances and the relative sensitivities of different species. Chronic toxicity tests are conducted over various time intervals, for example, 48 h, 96 h, 10 days, or 21 days, and measure the effects of substances ongrowth, development, reproduction, or behavior.

Most information on toxicity is based on the results of acute toxicity tests. Often this is the only information available on the effects of drilling fluids on marine organisms and thus is the information extrapolated for use in evaluating field situations in hazard assessments. Some of the limitations in extrapolating these tests should be recognized:

- Acute tests measure only lethality, not sublethal effects.
- They are not conducted over the course of organisms entire life stages or life cycles.
- They may not test species that are sensitive or commercially important.
- They require using such high concentrations of substances that they do not simulate the actual environmental exposure conditions, in which discharges may be diluted by 100 times within a few meters of the discharge pipe (see Chapter 3).

The method used to extrapolate from acute toxicity values to probable sublethal effects for a species is by using an application or safety factor. An application factor is the ratio of averaged acute and chronic values. Where no chronic test values are available for a species, a safety factor is used. For the results on drilling fluids, the safety (or application) factors that have been used range from $>0.1-.01$. In comparison, application factors for most toxicants, range from 0.01 to $>0.001$. Caution must be exercised in using these factors because the mechanism eliciting a sublethal toxic effect or an effect through chronic exposure may not be the same one producing the more easily measured acute toxicity. Ideally, application factors should be used in a hazard assessment only when the toxic responses in question result from the same mechanism; in practice this ideal is seldom attained. When assessing the actual ratios of acute to chronic toxicities, many being from sensitive species and life stages, the ratios range from 0.03 to 0.33, with the majority being towards the 0.33 end. This indicates that the normal safety factor is conservative.

Laboratory tests of the acute toxicities of drilling fluids have been conducted on over 70 species representing 5 major phyla. More than 36 percent of these tests have been conducted on organisms in larval and juvenile stages. Many of the tests of more sensitive species test were conducted on species in their early life stages: 48 percent of all shrimp tests, 43 percent of all decapod tests, 38 percent of all finfish tests, and 81 percent of all mysid tests (Neff et al., 1981; Petrazzuolo, 1983). Since 1980, the emphasis has been on testing the toxicities of drilling fluids on sensitive species and those in earlier developmental stages (eggs, larvae, and juveniles).

A range of 96-h LC50 values covering three orders of magnitude ($10^2$ to $10^5$ ppm) has been reported in testing drilling fluids. Petrazzuolo (1981) concluded that because of the distribution of these values (92 percent are greater than $10^3$ ppm) even sensitive oceanic species are unlikely to exhibit lethal toxicities to fluids much below the lowest known 96-h LC50 value, 50 to 100 ppm. This conclusion may be challenged if the observed distribution of toxicities is a function of variable fluid toxicity and not the result of testing with a few extremely sensitive and many insensitive species. If it is found that a small number of fluids are much more toxic than others, then the factors contributing to their toxicity must be identified.

It has been argued that the results of bioassays with intertidal, estuarine, and nearshore organisms should not be extrapolated to predict the effects of fluids on offshore species. The first groups of organisms often survive better any rapid changes in temperature and salinity, as well as the rigors of collection, transport, and being held in aquaria. It has been argued that these characteristics reflect these species' insensitivity to chemical pollutants. Recent studies indicate, however, that at least some nearshore species are as sensitive as those of similar morphology found offshore. For example, LC50 values for two copepods, estuarine *Acartia tonsa* and oceanic *Centropages typicus*, were similar in tests with several drilling fluids, even though the second species was much more difficult to keep in the laboratory (New England Aquarium, 1981). Moreover, a number of the species listed in Table 20 are found both near shore and on the outer continental shelf. These include the ocean scallop *Placopecten magellanicus* wich constitutes a commercial fishery on Georges Bank but which is also found along the coast of Maine. Other species with a similar range of habitat that have been the subject of bioassays include the bat star fish *Patiria miniata* (Chaffee and Spies, 1982), the cancer crab *Cancer borealis* (Gerber et al., 1981), and various species of echinoderms (embryos) (Crawford and Gates, 1981).

The design of some laboratory toxicity tests has also been criticized. For example, the EPA/COE method for assessing the toxicity of liquid, suspended particulate, and solid phases of drilling fluids at 1:4 dilution does not realistically separate the components of treated drilling fluids. The high ratio of fluid to seawater in this dilution (as compared to that in field conditions) results in a suspended particulate phase that is frequently unsuitable for testing the toxicity of the fluid. Thus, this method of preparing the bioassay mixture may confuse the estimated toxicity value of the drilling fluid or drilling-fluid fraction. Other problems with current bioassay techniques include their difficulty in filtering the suspended particulate phase (excess solids stay in suspension) and the opaqueness of their test solutions (especially of chrome lignosulfonate fluids), which makes conducting bioassay observations difficult. Small copepods have even been observed mired in layers of settled drilling-fluid solids (New England Aquarium, 1981), a situation unlikely to occur in nature.

The sophistication of toxicity tests of drilling fluids has improved in recent years. A growing body of data describes the sublethal effects of drilling fluids, effects including the abnormal development of mollusc larvae and embryos (Neff, 1980; New England Aquarium, 1981), decreased growth rates in mysid shrimp (Carr et al., 1980) and sea scallops (Gerber et al., 1981), and depressed feeding in adult lobsters Homarus americanus (Derby and Atema, 1981). Sublethal tests often examine organisms in critical life stages, and when they are properly designed their results allow a more realistic evaluation of the hazards posed by drilling fluids. In spite of the improvement that the tests here cited represent, the range of concentrations and the exposure durations they use may result in longer exposures than those occurring in the field.

The biological effects of discharged materials may also be assessed through microcosm studies of benthic larval recruitment (see Chapter 4). The resettlement of natural larval populations to defaunated sediment that has been mixed with or covered with drilling fluid in these microcosms may to some degree simulate the development of benthic communities following a drilling operation. Still, two factors limit the extrapolation of results from these tests to other regions. First, the effects on organisms of grain size, sediment chemistry, and other physical, chemical, and microbiological factors in the sediments have not always been isolated in experimental designs. Second, the results of these experiments are limited in that they apply only to larval populations.

In summary, laboratory toxicity testing has been useful in gauging the relative toxicities of drilling-fluid suspensions and will continue to be useful in screening drilling-fluid additives and in attempting to understand the mechanisms of toxicity and sublethal effects and the effects of short-term exposures. Given the data on the fates of drilling fluids in the field, expecially on rapid plume dispersion, and the available results of acute toxicity tests, as well as the inherent limitations in extrapolating from laboratory results, additional acute toxicity testing is unlikely to improve predictions of drilling fluids' effects on organisms in the water column. Laboratory tests have not realistically simulated the exposure conditions experienced by benthic organisms.

## BIOACCUMULATION

The bioaccumulation of metals from drilling fluids and cuttings has been addressed by both laboratory and field studies. In laboratory exposures to drilling-fluid components and in field situations both barium and chromium have been found to accumulate beyond levels in control organisms. Chronic ingestion of drilling-fluid solids by deposit-feeding organisms should be investigated further, since particulate metals may be accumulated under these circumstances (Liss et al., 1980). Such studies should consider that undigested solids would be eliminated from the digestive tract and should attempt to distinguish between metals that are nonspecifically bound to macromolecules (e.g., to enzymes and nucleic acids) and those associated with intracellular ligands, like metallothionein, or sequestered in membrane-bound vesicles and thus effectively detoxified (Jenkins and Brown,

1982). Bioaccumulation by organisms in the water column has not been examined directly. Their exposure to water-soluble phases is usually of short duration, as the fluid plume disperses rapidly, and thus bioaccumulation in these circumstances is unlikely.

## THE VARIABILITY OF DRILLING FLUIDS

It is important that a wide range of drilling fluids be evaluated in a comprehensive testing program. The cooperative program between the Petroleum Equipment Suppliers Association (PESA) and EPA to obtain samples of used drilling fluids for toxicity testing, which relied on random samples, gave some much-needed breadth to the data base on drilling-fluid composition and toxicity. Random sampling is essential in such a program if the results are to be credible. Complete documentation of the samples, detailing their source, the method used to obtain them, and their components and components concentrations are also needed to allow informed interpretation of chemical and toxicological testing.

It is clear from a review of the literature (Table 20, Chapter 4,) that the toxicities of drilling fluids to marine fauna vary up to three orders of magnitude even though the major constituents do not vary greatly. Toxic substances are added to some drilling fluids. Hexavalent chromium may be added to aid deflocculation, which it accomplishes mainly by extending the thinning ability of chrome lignosulfonate (Knox, 1978). Lubricants, such as diesel fuel, may be added to reduce torque along the drill string, particularly when drilling deviated (inclined) holes. The drilling fluids found to be relatively toxic include those from a well drilled in Mobile Bay in 1979 under a "no discharge" stipulation (Rubinstein et al., 1980) to which both hexavalent chromium and diesel fuel were added. These additives and their degradation products are probably the principal toxic agents in drilling fluids.

The purity of unrefined barite varies substantially; barite mined from vein displacement deposits may contain concentrations of lead and zinc sulfides in excess of 1 g/kg (Kramer et al., 1980). Because the concentrations of lead and zinc in deep ocean water are below $10^{-7}$ g/kg (Bruland, 1980; Patterson, 1974), discharges of drilling fluid weighted with barite will cause a temporary increase of these metals in the water column. These sulfide minerals dissolve slowly, however, so increases in their dissolved concentrations are difficult to detect.

## FIELD STUDIES OF THE FATES AND EFFECTS OF DRILLING FLUIDS

Direct measurement of the fates and effects of drilling fluids in the field is inherently more rigorous than a hazard assessment that relies on models to predict such fates along with laboratory measures of toxicity. On the other hand, field studies suffer the problem of site-specificity and that of distinguishing genuine effects from natural variability. Furthermore, even when effects are documented in the field, it is generally difficult to determine the significance of their geographic extent or duration for the ecosystem or for man. Final appraisals thus become highly subjective. Finally, field assessments suffer the inherent limitations of retrospective studies.

Field determinations of environmental fates have been of two types: monitoring dissolved and particulate concentrations of drilling-fluid constituents during discharge, and monitoring the concentrations of drilling-fluid constituents in bottom sediments and organisms after discharge.

In studies of the water column, the visible or detectable plume of suspended particulate matter has sometimes been sampled for potential toxicants. Several recent studies of plume dispersion employed *in situ* transmissometry techniques (Ayers et al., 1980; Ray and Meek, 1980) or acoustical techniques (Proni and Trefry, 1981) to sample areas where concentrations of suspended matter were highest. In near-field dispersion, the advection and dispersion of dissolved and suspended phases in the surface plume should be qualitatively similar; greatly underestimating dissolved concentrations in the water column seems unlikely. Plume dispersion studies have been conducted using dye releases, transmissometer profiling, or discrete water grabs for a variety of discharge rates (bulk and continuous), water depths (23 to 120 m), and current speeds (16 to 120 cm/s). The results of these empirical studies support the theoretical models (see Chapter 3) and indicate that the soluble phase is diluted by at least $10^4$ within 1 h after discharge. Of course, the distance from the discharge point at which a particular dilution is reached will vary depending on current velocity, and time from discharge is generally a better predictor of dilution. Apparent dispersion of the suspended particulate phase is at least an order of magnitude greater, because most of the discharged drilling fluids (probably more than 90 percent, especially of bulk discharges) and essentially all of the cuttings sink to the seabed within a short distance of the discharge (in depths less than about 125 m--at greater depths the discharges reach neutral buoyancy before encountering the seafloor).

With the dispersion of potentially soluble toxicants by a factor of $10^4$ within 1 h of discharge (corresponding usually to a spatial extent of about 1,000 m), toxic responses in this zone should be anticipated only if short-term exposures of several hours to the substance discharged produced EC50 values in the 100 ppm (v/v) range. This conclusion is strongly supported by the plume effects model of the EPA Adaptive Environmental Assessment Workshop (Auble et al., 1982) and by Petrazzuolo's (1981) dispersion toxicity models. The results of recent analyses of sublethal effects in organisms at critical life stages indicate that discharges of drilling fluids would usually not approach such values. This conclusion cannot yet be extrapolated with confidence to shallow-water environments (< 10 m) and embayments where dispersion may not be as rapid, although recent dispersion measurements from the Beaufort Sea (Nortec, 1983) suggest similar dispersion in shallow water.

Direct field surveys of the effects on planktonic or nektonic organisms have not been attempted and are probably not feasible given natural and sampling variability and the turbulent mixing of potentially affected and unaffected populations, and in light of some of the conclusions of prior hazard assessments.

A significant shortcoming in understanding the long-term effects of drilling discharges concerns the fate of particulate components once they reach the seabed. An important factor in determining the fate of pollutants is the resuspensive transport of sediments that tends to dilute and disperse particulate contaminants. Transport depends on the hydrodynamic regime of an environment. While most of the fluids discharged on Tanner Bank (Meek and Ray, 1980) and in Lower Cook Inlet (Dames and Moore, 1978) settled to the seabed rapidly, no accumulation of contaminants in bottom sediments was observed because strong currents resuspended and dispersed the discharged material. On the other hand, gradients in barium concentration persisted in the more quiescent benthic environments of the Gulf of Mexico (Boothe and Presley, 1983; Gettleson and Laird, 1980; Trocine et al., 1981) and the shelf break of the Middle Atlantic Bight (EG&G Environmental Consultants, 1982). Within the Gulf of Mexico, Boothe and Presley (1983) also found that the degree to which they could account for the total barium discharged in sediments surrounding an exploratory rig was directly related to water depth. In 13 m of the water, only 5.4 percent of the barium discharged could be accounted for in sediments within 3 km of the rig. At another exploratory well in 13 m depth 84 percent of the barium could be accounted for within a similar radius and 9.6 percent within a radius of 500 m. At a production platform where 25 wells had been drilled in 79 m of water, only 11.6 percent of the barium was found within 500 m as compared to 1.5 percent at a production platform in 34 m of water (see Table 16).

In most reports, elevated levels of major drilling-fluid components like barium were confined to an area within 1 km of the discharge point. Care must be taken in interpreting such results because concentrations of drilling fluids can be diluted beyond detection in samples that include the upper two cm of surficial sediments. In one study that sampled only the top 1 cm of sediment, barium concentrations were three times ambient concentrations 1.9 km from the well (Trocine et al., 1981). Surface layer contamination may pose elevated exposure conditions to those benthic organisms that feed at the sediment-water interface. With time, sediment contaminants will disperse horizontally and also be vertically mixed in sediments. Boothe and Presley (1983) found incorporation of barium to at least 15 cm near a production platform more than 5 years after drilling had ceased.

Few studies have attempted to measure the temporal extent of benthic changes. The effects of discharges on the benthos depends greatly on how quickly the community recovers, not only in total density and biomass, but also in the composition and structure of the community. Populations of larger, deeper burrowing benthic organisms, which contribute to the geochemical structure of sediments and to other features of the benthos by their feeding, burrowing, and respiration, recover more slowly than small surface dwellers (Boesch and Rosenberg, 1981; Rhoads et al., 1978). Particularly in outer shelf habitats, important species may have populations dominated by individuals several years old, so substantial time is required to reestablish the natural age structure in the community, even given rapid recolonization. Gillmor et al. (1981), in the only study sampling the benthos 1 year after drilling-fluid discharges ceased, found reduction in the density of the ophiuroid Amphioplus macilentus at the site of an exploratory well on the outer middle Atlantic shelf. Amphioplus is an important burrower in the benthic community in this habitat and is probably long lived, as indicated by its persistent populations and community structure. Amphioplus also showed depressed recruitment in affected areas.

Drilling-fluid discharges may more greatly damage the ecosystem if the spatial extent of their effects transcends those observed through chemical analyses of sediments, or if their effects are long lasting, because of slow recovery of communities or habitat modification (or both). The postdepositional fates of drilling fluids and the recovery of altered communities are the processes for which data are most limited and predictions most tenuous. Hydrodynamic regimes, including tidal and nontidal currents and wave-induced orbital water movements, obviously vary from one region to another and across the continental shelf. Furthermore, information on the resilience of benthic communities suggests that recovery rates from complete annihilation vary from weeks in shallow-water communities (that are frequently disturbed by nature), to several months or years for continental shelf communities, and to many years on the continental slope and in the deep sea (Boesch and Rosenberg, 1981). The variability in dispersion in the benthic boundary layer, the resistance of biota to physical and toxic effects, and the resilience of communities in different continental shelf environments all need to be taken into account in assessing benthic effects (Auble et. al., 1982; Petrazzuolo, 1981).

Concern about the sensitivity of hard-substrate epibiota to the physical and toxic effects of drilling fluids has prompted special studies and regulatory restrictions, such as those on the Flower Garden reefs and Tanner Bank. This concern is often translated into treating all areas where hard-substrate epibiota exist (such as reefs, rocky outcrops, and canyon heads) as "biologically sensitive areas" in environmental impact statements and when applying lease stipulations or permit requirements. A characteristic feature of hard-substrate communities is a lack of sediment cover. The absence of sediments allows the colonization and proliferation of colonial or solitary epibiota on the hard substrate, which enhances the structure of the habitat and affords habitation to a variety of motile animals seeking refuge or food. The lack of sediment cover may result from a dynamic physical regime that sweeps sediments away or from the lack of a source of fine sediments for deposition. In the first case, drilling fluids dumped or advected into the habitat may not be deposited or accumulate. Thus, despite the sensitivity of its biota, the habitat is not very susceptible to harmful accumulation of drilling fluids. Concern should be directed to hard-substrate communities in more quiescent habitats. Organisms in these communities may be sensitive to nearby discharges of drilling fluids if the fluids' rate of accumulation exceeds the organisms' ability to remove settling material or if the material is toxic. The sensitivity of hard-substrate communities should be evaluated in light of their potential exposure to drilling fluids and cuttings rather than through assuming categorically that hard-substrate habitats are "biologically sensitive."

## EXTRAPOLATION OF RESULTS

A common concern in using research results is whether they can be extrapolated to a particular case. Such extrapolation includes that from laboratory and other experimental results to the natural environment and also that from one environment or geographic area to another. Marine ecosystems on the OCS clearly vary in their sensitivities to anthropogenic stress, and caution is therefore advisable in extrapolating observations from one region to another. On the other hand, to dismiss all research results not obtained directly from the environment analyzed may amount to ignoring valuable data. Most important in extrapolating results are considering the kind of physicochemical processes affecting the fates of contaminants and the resistance and resilience of affected communities.

There are generally adequate data to predict the fates of dissolved and suspended drilling-fluid components in the water column in different OCS environments. Such data are not necessary available on the long-term fates of deposited materials. The general model of a continuum of marine environments from those relatively dynamic (for example, in Cook Inlet or on the inner continental shelf) to those quiescent (for example, in the Gulf of Mexico or on the outer shelf) is adequate to conceptualize these differences but not to quantify them. The emerging theories of the dynamics of the bottom boundary layer, including their treatment of the effects of surface roughness, biological processes, and heterogeneity of grain size, will be required in making sound quantitative extrapolations.

Laboratory experiments have shown no clear differences in the relative sensitivities of organisms from different geographic regions to drilling fluids. Given that drilling fluids' most profound effects will likely be on the benthos, this suggests that relative biological susceptibility will be determined primarily by the fates of deposited materials and by a biological community's ability to recover (resilience). The resilience of benthic communities varies significantly. It can be predicted approximately (Boesch and Rosenberg, 1981), but not precisely, for most hard- and soft-substrate communities.

In summary, because of the lack of quantitative models of the fates of deposited materials and biological resistance and resilience, the extrapolation of results from one environment or geographic area to another can presently be only qualitative.

## LONG-TERM FATES AND EFFECTS

Most information available on the fates and effects of drilling fluids and cuttings comes from studies of single exploratory wells. The long-term effects of drilling discharges clearly may be greater when the discharges are made during oil and gas field development, when scores of wells may be drilled from a single platform or within a lease block. In drilling multiple development wells from a single platform (see Chapter 2) the volume of drilling fluids discharged per well is substantially less, but the mass loading within an area and the duration of discharges are greater. For example, extensive drilling in the Gulf of Mexico OCS has gone on for 30 years, with an estimated current annual discharge of drilling-fluid solids of 1.6 million t (Gianessi and Arnold, 1982), while in the first few years of exploring a frontier area between 10,000 and 100,000 t of drilling-fluid solids may be discharged.

Several studies have addressed the long-term effects of oil and gas development and production in the Gulf of Mexico, notably, the Offshore Ecology Investigation (Ward et al., 1979), the Buccaneer Field Study (Middleditch, 1981), and the Central Gulf Platform Study (Bedinger, 1981). The Offshore Ecology Investigation conducted in 1972 to 1973 contributed little to understanding the fates of contaminants resulting from petroleum development because it failed to measure the contaminants in bottom sediments adequately. The other studies examined the distribution of potential sediment contaminants (trace metals and hydrocarbons) along gradients from discharge points and compared these data to those obtained from presumably uncontaminated areas. Of course, there are many sources of contaminants other than drilling fluids and cuttings related to the platform and its operation. These include corrosion of materials, produced water discharges, sacrificial anodes, domestic wastes, vessel discharges, and chemicals used in operating the platform. At most of the platforms studied, many trace metals were elevated (including mercury, lead, copper, and zinc) in surface sediments compared to areas removed from development and subsurface sediments, and were found in gradients around the platform (Tillery, 1980a,b; Tillery et al., 1981; Boothe and Presley, 1983). Interestingly, Boothe and Presley (1983) found no evidence of elevated concentrations of chromium, the only other metal for which drilling fluids are a likely source. Barium was the only trace metal elevated around development and production platforms whose most likely source is drilling fluids, although it can also be present in formation waters at higher concentrations than in seawater.

Barium concentrations in surface sediments within 100 to 200 m of Buccaneer Field platforms (17-22 m water depth) were higher than those in subsurface sediments and surface sediments in undeveloped areas of the south Texas continental shelf (Anderson et al., 1981). Decreasing barium concentration gradients were also observed with increasing distance from the Buccaneer Field platforms (Tillery, 1980a,b), from many of those off Louisiana sampled during the Central Gulf Platform Study (Tillery et al., 1981), and from all of the six rigs considered in Boothe and Presley's recent study (1983) of exploration, development, and production sites. Near Buccaneer Field plataforms, Anderson et al. (1981) also noted the presence of bentonite clay, atypical of local marine sediments and possibly originating from drilling fluids. On the other hand, Boothe and Presley (1983) were unable to distinguish any bentonite clay that could have come from drilling fluids from the similar montmorillonite clays naturally present in Gulf of Mexico sediment using simple bulk x-ray diffraction. It is also noteworthy that one study found barium concentration gradients around several platforms in the Mississippi delta region (Tillery et al., 1981), despite the authors' assertion that the influence of the Mississippi River masks platform-related contamination.

The localized effects (within 100 m) on benthic communities observed in the Buccaneer Field (Harper et al., 1981) cannot be related unequivocably to drilling discharges because of the other contaminant sources and physical effects associated with the platforms (for example, seabed scour). However, the barium tracer data are significant in that 10 or more years had elapsed since active drilling to the time of sampling at some of these platforms. With regard to the fates of drilling fluids these data suggest that detectable contamination of bottom sediments, which may or may not have biological effects, may persist under some OCS sedimentary regimes for years and perhaps decades. The results of these Gulf of Mexico studies are, at present, insufficient to quantify the long-term effects of drilling discharges from large-scale offshore oil and gas development in the Gulf of Mexico and to extrapolate elsewhere. This is in part because of the paucity of long-term observations and because of the difficulties in separating the effects of drilling discharges from those of other activities.

Given the limitations of these Gulf of Mexico studies, what do the assessments of single exploratory wells suggest about the long-term effects of more massive drilling discharges during field development? Are the operative transport processes for the two kinds of discharges quantitatively similar? Do greater discharges result in greater, more extensive or more persistent contamination? Are the effects of greater discharges simply additive or are they different from those of discharges from single wells?

These questions cannot be answered with absolute certainty. Even so, the results of studies of exploratory wells should be pertinent to the assessment of long-term effects. In the water column, the elevated concentrations of contaminants and their effects should be very small and transient. Documented effects of long-term discharges on the benthos are areally limited and transient. The fates of deposited materials are more strongly influenced by the dispersal regime (for example, as determined by water depth) than by any other factor. Contamination of bottom sediments from multiple wells appears to be less than simply additive (Boothe and Presley, 1983). Despite these conditions, the existence of subtle effects caused by contamination over broad areas in heavily developed environments cannot be ruled out.

## OTHER INFORMATION

Information available on the fates and effects of drilling fluids and cuttings shows that the effects of an individual discharge are likely to be limited in extent and primarily confined to the benthos. Research to date indicates that the environmental risks of discharges from exploratory drilling to most OCS communities are small. No additional research is needed on the fates and effects of drilling fluids in the water column in open water where rapid mixing is likely.

These conclusions do not hold for shallow-water, low-energy environments like estuaries or embayments or for some areas under ice in the Arctic. If additional information on these topics is sought, it should be on the fates and effects of materials discharged in the development of production wells and on particularly susceptible environments.

The above issues are general ones applying to the fates and effects of all human inputs into the coastal ocean. Better understanding of the potential contamination of nearshore environments would probably best be attained by generic studies of the processes that determine the ecological effects of foreign substances in these environments.

## Transport and Transformation

There is little information on the dispersion of drilling fluids and cuttings in the bottom boundary layer. The vast majority of drilling fluids and cuttings discharged into the water column settle in little time near the discharged point. Resuspension and tractive (bed load) transport determine the persistence, dispersion, and ultimate fates of contaminants associated with this particulate material. The tidal and mean currents and wave climates that affect these sediment transport processes vary widely among OCS areas. Such differences in physical regimes can be clearly seen in comparisons of sediment contamination in Cook Inlet, Tanner Bank, the Middle Atlantic Bight, and the Gulf of Mexico (see Chapter 3). Significant advances have been made recently in measuring sediment transport in the bottom boundary layer and in modeling the interactive effects of waves and currents (Grant and Madsen, 1982). This emerging technology and theory should be applied in any development of predictive models on the fates of the sedimentary fractions of drilling fluids and cuttings, if the models are to be relevant to the variety of erosional and depositional OCS environments. Such understanding is important in assessing the fate of contaminants from long-term oil and gas development. Predicting the physical fates of contaminants is also important in judging the susceptibility of sensitive or valuable environments, such as hard-bottom banks and the estuaries adjacent to nearshore discharges.

The bioaccumulation as well as the transport of trace metals in marine discharges has been extensively studied. The data on the subject suggest that bioavailability is related to the chemical activity of the metal ions. The bioaccumulation of barium in marine bivalves, however, appears to be related to the loading of particulate $BaSO_4$ and not just to the concentration of barium ions in the environment. The mechanism of the uptake of barite particles is not well understood, nor are the composition, transformation, and bioavailability of some of the organic additives in used drilling fluids. Petroleum hydrocarbons are included in this category since they are often found in water-based drilling fluids.

## Effects

The biological effects of drilling discharges are restricted primarily to the benthos. The conditions of benthic exposure and toxicity are only partially understood. These conditions include the rates and routes of bioavailability in benthic organisms to contaminants in resuspended particles and interstitial waters. Documenting these conditions would require coordinated laboratory and experimental field approaches because of the difficulty in duplicating exposure conditions in conventional bioassays and the need to assess "sublethal" effects in light of their significance for the organism's life and for the survival of populations. Another part of the analysis would describe the relative resistance of benthic communities to the physical and chemical effects of sediment contamination from anthropogenic inputs, and the resilience (or speed of recovery) of affected communities.

The development of sensitive and reproducible methods for testing toxicity is required to standardize assessments of drilling-fluid components and additives.

## Resource Management

Like most environmental research, that on drilling-fluid discharges in the marine environment should be closely coordinated with resource management. Particular regulatory needs in this field include specifying operational alternatives for the composition and discharge of drilling fluids (especially with regard to additives, including diesel fuel) and standardizing toxicity tests to screen drilling fluids and their components.

## REFERENCES

Anderson, J.B., R.B. Wheeler, and R.S. Schwarzer. 1981. Sedimentology and geochemistry of recent sediments. In: B.S. Middleditch (ed.), Environmental Effects of Offshore Oil Production. New York: Plenum Press. Pp. 59-67.

Armstrong, R.S. 1981. Transport and dispersion of potential contaminants. In: B.S. Middleditch (ed.), Environmental Effects of Offshore Oil Production. New York: Plenum Press. Pp. 403-420.

Auble, G.T., A.K. Andrews, R.A. Ellison, D.B. Hamilton, R.A. Johnson, J.E. Roelle, and D.R. Marmorek. 1982. Results of an adaptive environmental assessment modeling workshop concerning potential impacts of drilling muds and cuttings on the marine environment. Office of Biological Services, U.S. Fish and Wildlife Service, Fort Collins, Colo.

Ayers, R.C., Jr., T.C. Sauer, Jr., D.O. Steubner, and R.P. Meek. 1980. An environmental study to assess the effect of drilling fluids on water quality parameters during high rate, high volume discharges to the ocean. In: Proceedings of a Symposium on Research on Environmental Fate and Effects of Drilling Fluids and Cuttings. Washington, D.C.: Courtesy Associates. Pp. 351-391.

Bayne, B.L. , D.A. Brown, F.L. Harrison, and P.P. Yevich. 1980. Mussel health. In: E. Goldberg (ed.), International Mussel Watch, Report on a Workshop at Barcelona, Spain, December 1978. Washington, D.C.: National Academy of Sciences. Pp. 196-235.

Bedinger, C.A., Jr. (ed.). 1981. Ecological investigations of petroleum production platforms in the central Gulf of Mexico. Report to Bureau of Land Management. Contract AA551-CT8-17. Southwest Research Institute, San Antonio, Tex.

Boesch, D.F., and R. Rosenberg. 1981. Response to stress in marine benthic communities. In: G.W. Barrett and R. Rosenberg (eds.), Stress Effects on Natural Ecosystems. New York: John Wiley & Sons. Pp. 179-200.

Boothe, P.N., and B.J. Presley. 1983. Distribution and behavior of drilling fluid and cuttings around Gulf of Mexico drill sites. Draft final report. API Project No. 243. American Petroleum Institute, Washington, D.C.

Bruland, K.W. 1980. Oceanographic distributions of cadmium, zinc, nickel and copper in the north Pacific. Earth Planet. Sci. Lett. 47:176-198.

Carls, M.G., and S.D. Rice. 1980. Toxicity of oil well drilling muds to Alaskan larval shrimp and crabs. Research unit 72. Final report. Project No. R7120822. Outer Continental Shelf Energy Assessment Program, Bureau of Land Management, U.S. Department of the Interior. 29 pp.

Carr, R.S., L.A. Reitsema, and J.M. Neff. 1980. Influence of a used chrome-lignosulfonate drilling mud on the survival, respiration, growth, and feeding activity of the opossum shrimp *Mysidopsis almyra*. In: Proceedings of a Symposium on Research on Environmental Fate and Effects of Drilling Fluids and Cuttings. Washington, D.C.: Courtesy Associates. Pp. 944-963.

Dames & Moore, Inc. 1978. Drilling fluid dispersion and biological effects study for the lower Cook Inlet C.O.S.T. well. Report submitted to Atlantic Richfield Co. Dames & Moore, Anchorage, Alaska. 309 pp.

Dames & Moore, Inc. 1981. Fate and effects of drilling fluids and cuttings discharges in lower Cook Inlet, Alaska and on Georges Bank. Final report to National Oceanic and Atmospheric Administration and Bureau of Land Management. 336 pp., 3 appendixes.

Derby, C.D., and J. Atema. 1981. Influence of drilling muds on the primary chemosensory neurons in walking legs of the lobster, *Homarus americanus*. Can. J. Fish. Aquat. Sci. 38:268-274.

EG&G Environmental Consultants. 1982. A study of environmental effects of exploratory drilling on the mid-Atlantic Outer Continental Shelf--final report of the Block 684 Monitoring Program. EG&G Environmental Consultants, Waltham, Mass. Available from Offshore Operators Committee, Environmental Subcommittee, P.O. Box 50751, New Orleans, LA 70150.

Gettleson, D.A., and C.E. Laird. 1980. Benthic barium in the vicinity of six drill sites in the Gulf of Mexico. In: Proceedings of a Symposium on Research on Environmental Fate and Effects of Drilling Fluids and Cuttings. Washington, D.C.: Courtesy Associates. Pp. 739-788.

Gianessi, L.P., and F.D. Arnold. 1982. The discharges of water pollutants from oil and gas exploration and production activities in the Gulf of Mexico region. Draft report to National Oceanic and Atmospheric Administration, Contract NA-80-SAC-00793. Resources for the Future, Washington, D.C.

Gilfillan, E.S., R.P. Gerber, S.A. Hanson, and D.S. Page. 1981. Effects of various admixtures of used drilling mud on the development of a boreal soft bottom community (unpublished manuscript). American Petroleum Institute, Washington, D.C. 17 pp.

Gillmor, R.B., C.A. Menzie, G.M. Mariani, D.R. Levin, R.C. Ayers, Jr., and T.C. Sauer, Jr. 1981. Effects of exploratory drilling discharges on the benthic environment in the middle Atlantic OCS: biological results of a one-year post-drilling survey. In: Proceedings, 3rd International Ocean Disposal Symposium, October 12-16, Woods Hole, Mass.

Grant, W.D., and O.S. Madsen. 1982. Combined wave and current interaction with a rough bottom. J. Geophys. Res. 84:1797-1808.

Harper, D.E., Jr., D.L. Potts, R.R. Salzer, R.J. Case, R.L. Jaschek, and C.M. Walker. 1981. Distribution and abundance of macrobenthic and merobenthic organisms. In: B.S. Middleditch (ed.), Environmental Effects of Offshore Oil Production. New York: Plenum Press. Pp. 133-177.

Jenkins, K.D., and D.A. Brown. In press. Determining the biological significance of contaminant bioaccumulation. In: H. White (ed.), Proceedings of the Workshop on Meaningful Measures of Marine Pollution Effects. Pensacola, Fla., April 1982. College Park, Md.: University of Maryland Sea Grant Program.

Knox, F. 1978. The behavior of ferrochrome lignosulfonate in natural waters. Masters Thesis. Massachusetts Institute of Technology, Cambridge, Mass. 65 pp.

Koh, R.C.Y., and Y.C. Chan. 1973. Mathematical model for barged ocean disposal wastes. EPA-660/2-73-029. Pacific Northwest Environmental Research Laboratory, Environmental Protection Agency. Corvallis, Oreg.

Kramer, J.R., H.D. Grundy, and L.G. Hammer. 1980. Occurrence and solubility of trace metals in barite for ocean drilling operations. In: Proceedings of a Symposium on Research on Environmental Fate and Effects of Drilling Fluids and Cuttings. Washington, D.C.: Courtesy Associates. Pp. 789-798.

Liss, R.G., F. Knox, D. Wayne, and T.R. Gilbert. 1980. Availability of trace elements in drilling fluids discharged to the marine environment. In: Proceedings of a Symposium on Research on Environmental Fate and Effects of Drilling Fluids and Cuttings. Washington, D.C.: Courtesy Associates. Pp. 691-722.

Meek, R.P., and J.P. Ray. 1980. Induced sedimentation, accumulation, and transport resulting from exploratory drilling discharges of drilling fluids and cuttings. In: Proceedings of a Symposium on Research on Environmental Fate and Effects of Drilling Fluids and Cuttings. Washington, D.C.: Courtesy Associates. Pp. 259-284.

Middleditch, B.S. (ed.). 1981. Environmental Effects of Offshore Oil Production. New York: Plenum Press. 446 pp.

Neff, J.M. 1980. Effects of used drilling muds on benthic marine animals. Publ. No. 4330. American Petroleum Institute, Washington, D.C. 31 pp.

Neff, J.M. 1981. Fate and biological effects of oil well drilling fluids in the marine environment: a literature review. Final technical report to U.S. Environmental Agency, Gulf Breeze, Fla. 151 pp., 2 appendices.

Neff, J.M., R.S. Carr, and W.L. McCulloch. 1981. Acute toxicity of a used chrome lignosulfonate drilling mud to several species of marine invertebrates. Mar. Environ. Res. 4:251-266.

New England Aquarium. H.E. Edgerton Research Laboratory. 1981. A study of the impact of discharged drilling fluids on the Georges Bank environment. Progress Report No. 2 to the U.S. Environmental Protection Agency, Gulf Breeze, Fla. 98 pp.

Page, D.S., B.T. Page, J.R. Hotham, E.S. Gilfillan, and R.P. Gerber. 1980. Bioavailability of toxic constituents of used drilling muds. In: Proceedings of a Symposium on Research on Environmental Fate and Effects of Drilling Fluids and Cuttings. Washington, D.C.: Courtesy Associates. Pp. 984-996.

Patterson, C.C. 1974. Lead in seawater. Science 183:553-554.

Petrazzuolo, G. 1981. Preliminary report: an environmental assessment of drilling fluids and cuttings released onto the Outer Continental Shelf from the Gulf of Mexico. Vol. I: Technical assessment. Vol. II: Tables, figures and Appendix A. Draft report prepared for Industrial Permits Branch, Office of Water Enforcement and Ocean Programs Branch, Office of Waster and Waste Management, U.S. Environmental Protection Agency, Washington, D.C.

Petrazzuolo, G. 1983. Environmental assesment: drilling fluids and cuttings released onto the OCS. Draft final technical support document. Submitted to the Office of Water Enforcement and Permits, U.S. Environmental Protection Agency, Washington, D.C.

Proni, J., and J.A. Trefry. 1981. Drilling fluid discharge in the Gulf of Mexico. Paper presented at Third International Ocean Disposal Symposium, Woods Hole, Mass. October 12-16.

Rhoads, D.C., P.L. McCall, and J.Y. Yingst. 1978. Disturbance and production of the estuarine sea Ffoor. Am Sci. 66:577-586.

Rubinstein, N.I., R. Rigby, and C.N. D'Asaro. 1980. Acute and sublethal effects of whole used drilling fluids on representative estuarine organisms. In: Proceedings of a Symposium on Research on Environmental Fate and Effects of Drilling Fluids and Cuttings. Washington, D.C.: Courtesy Associates. Pp. 828-846.

Tagatz, M.E., J.M. Ivey, H.K. Lehman, and J.L. Oglesby. 1978. Effects of lignosulfonate-type drilling mud on development of experimental estuarine macrobenthic communities. Northeast. Gulf Sci. 2:25-42.

Tillery, J.B. 1980a. Trace metals. Vol. VIII in W.B. Jackson and E.P. Wilkens (eds.), Environmental assessment of Buccaneer gas and oil field in the northwest Gulf of Mexico, 1978-1979. NOAA/NMFS Annual Report to EPA. NOAA Tech. Memorandum NMFS-SEFC-42. 93 pp.

Tillery, J.B. 1980b. Trace metals. Vol. VI in W.B. Jackson and E.P. Wilkens (eds.), Environmental assessment of Buccaneer gas and oil field in the northwestern Gulf of Mexico, 1978-1979. NOAA/NMFS Annual Report to EPA. NOAA Tech. Memorandum NMFS-SEFC-52. 39 pp.

Tillery, J.B., R.E. Thomas, and H.L. Windom. 1981. Trace metal studies in sediment and fauna. In: C.A. Bedinger (ed.), Ecological Investigations of Petroleum Production Platforms in the Central Gulf of Mexico. Vol. I, Part 4. Report to Bureau of Land Management, Contract AA551-CT8-17. San Antonio, Tex.: Southwest Research Institute. Pp. 1-122.

Trocine, R.P., J.H. Trefry, and D.B. Meyer. 1981. Inorganic tracers of petroleum drilling fluid dispersion in the northwest Gulf of Mexico. Reprint extended abstract. Div. Environ. Chem. ACS Meeting, Atlanta, Ga. March-April 1981.

Ward, C.H., M.E. Bender and D.J. Reish (eds.). 1979. The offshore ecology investigation. Rice Univ. Stud. 65(4&5). 589 pp.

# 6 Alternative Operating Practices

In response to federal and state statutory requirements and concerns, a number of alternatives to simple overboard discharge have been employed or developed. Table 24 describes alternatives that have been required, and also others that have been developed or considered. This chapter briefly discusses these alternatives with regard to operations, cost, and risk.

## SHUNTING

Shunting refers to the discharge of drilling fluids and cuttings through a down pipe (shunt pipe) to a predetermined water depth. Shunting has been required for some OCS wells to reduce the exposure of organisms in the water column or to transport discharged material to the bottom boundary layer to reduce the exposure of sensitive communities on topographic rises. Shunting to the water column probably has little effect on dispersion. Where the bottom boundary layer is slower circulating than other water masses, shunting to the bottom can reduce the rate of dispersion.

While shunting systems can be designed for and operated in water of any depth, their costs and operating problems increase with the depth of the system, and also with the severity of the weather. Shunts have been used in the Gulf of Mexico in 100 m of water. The addition of a shunt system adds equipment and weight to the already-crowded drilling rig and another appendage below the water line in proximity to the marine riser and blowout control systems. If the shunt pipe were to swing loose because of heavy weather or damage it could collide with and damage subsea connections.

Shunting operations for one well in the Gulf of Mexico employing a 100-m shunt system on a jackup rig were estimated to cost about $107,000 (1982).[1] With proper care, such a system can be used

---

[1] James Gonders, Cities Service , July 1982, personal communication.

TABLE 24 Discharge Alternatives

| Alternative | Objective | Examples of Where Used | Examples of How Required |
|---|---|---|---|
| Shunting near surface | Minimize exposure of plankton | North Atlantic, mid-Atlantic | EPA permit, MMS stipulations |
| Shunting near bottom | Minimize exposure of coral reefs | Flower Garden Banks | EPA permit, Region VI |
| Dilution requirements, rate of discharge limitations | Reach greater dilution to limit harm to biota in the water column | Lower Cook Inlet, Georges Bank | EPA permits, regions X and I |
| Barging to land | Avoid ocean discharge | Alabama and California State Offshore Lands | State regulation |
| Barging to ocean dump site | Avoid discharge in coastal environment | -- | (Requires EPA-designated ocean dump site) |
| Disposal on ice | Take advantage of seasonal ice breakup to dissipate effluents | Beaufort Sea | Lease stipulation |
| Generic muds and approved additives | Limit toxicity of fluid | North Atlantic, mid-Atlantic, California, Alaska | EPA permits, regions I, II, IX, X |
| Alternate processing/ recycling/ reuse | Remove undesired components, minimize discharge | | |
| Incineration | Remove oil-contaminated cuttings | | |
| Injection | Reduce open-ocean discharge | | |

on several wells. Shunting is considered a relatively inexpensive discharge option.

## DILUTION REQUIREMENTS AND LIMITATIONS ON RATES OF DISCHARGE

Drilling fluids and cuttings can be (and normally are) discharged at or near the water surface. This practice results in a visible plume that may extend over several kilometers. The extent and duration of the plume depends on the energy of the ocean environment. In certain areas, notably Lower Cook Inlet, Alaska, predilution or certain discharge rates have been required in response to seasonal conditions.

The methods of predilution and maintaining certain discharge rates may require additional equipment, such as pumps and special pit gauges to monitor discharge rates. These special methods can affect cost and operations by affecting the duration of operations. Drilling must occasionally be stopped to complete a bulk discharge that is prolonged because of the high volumes of water required for predilution and the slow discharge rates allowed. This added time translates directly into cost for the operator. At other times, the necessity of completing a prolonged discharge may restrict the time available to move mobile rigs with respect to weather conditions and sea states. The failure to take advantage of good conditions for these activities can increase the risk of the operation. The data on dispersion presented in Chapter 3 indicates that such requirements for predilution and restrictive discharge rates are not justified in most OCS areas.

## OFFLOADING AND TRANSPORT FOR DISTANT DISCHARGE

Drilling discharges can be transported by barge or supply boat to an ocean or land disposal site. Ocean disposal requires the designation of a site in accordance with EPA ocean dumping regulations (40 CFR 220-230). Release at an ocean dump site would presumably take the discharge from a coastal environment for dispersal in deep water. Ocean disposal sites in the mid-Atlantic have been used for industrial wastes and could be used for drilling discharges as well, but no ocean dump sites have yet been used for these discharges. Land disposal also requires a suitable site, but disposal areas for industrial wastes on land are increasingly at a premium. Obtaining ocean dumping permits or disposing of drilling discharges on land adds to the cost of drilling operations.[2]

The ability to offload discharges for transport is directly related to sea states and weather conditions. Adverse conditions will prevent offloading as shown in Table 25.

---

[2] For example, one offshore operator paid $390,000 to barge drilling discharges from a well to shore (W. D. Fritz, Mobil Oil, personal communication, 1980) The landfill operator who received the wastes promptly sold them for fill dirt.

TABLE 25 Percentage of Time That Drilling Discharges Cannot Be Transferred to Barges or Supply Boats

| | Gulf of Mexico | Georges Bank |
|---|---|---|
| Offloading to barges (when seas exceed 1 m) | 20 | not feasible[a] |
| Offloading to supply boats (when seas exceed 3 m) | 2.2 | 12.4 |

[a]Seas exceed 1 m too much of the time to plan such operations.

SOURCE: Adapted from OOC, 1981.

During adverse weather or sea states that prevent the off loading of drilling discharges, drilling operations might have to be curtailed due to a lack of on-rig storage space, resulting in substantial additional costs.

Some units, such as small jack-up drilling rigs, have very little holding capacity, while other units, such as large semisubmersibles, have a greater holding capacity. Adequate holding capacity can lessen rig downtime in adverse conditions.

Offloading and transport increase the hazards of offshore drilling operations. Care must be taken to moor the receiving vessel to prevent damage to the drilling unit. The approach and mooring of the barge or vessel are constrained by weather and by the mooring arrangements of the drilling unit. The position of crane facilities on the drilling unit dictates the available loading points. The addition of anchoring systems required on a disposal barge further compounds the hazards. Once the transfer of the discharge has been made, additional risk is entailed in the transit of the transfer vessel to other areas for disposal and in the additional handling of the drilling discharges in disposal.

The cost of offloading and transport operations varies with the circumstances. For an 18,000-ft (5,490-m) well in the Gulf of Mexico, the cost of these operations has been estimated at $917,000 (OOC, 1981). Transport of discharges to shore from a comparable well drilled off Georges Bank would cost about $3.29 million; to an ocean dump site $3.02 million (OOC, 1981). Such costs range from 10 to 20 percent of the cost of the well.

## OTHER TRANSPORT TECHNIQUES

Other techniques suggested for transporting discharges include a pipeline to an ocean dump site and the use of a monobuoy for loading discharges onto a barge or vessel at a distance from the drilling unit. These arrangements have been reviewed (OOC, 1981), but are not considered here because of their limited applications and higher costs.

## DISPOSAL ON ICE

One discharge alternative has been suggested for use in the Beaufort Sea and other areas where sea ice is present for parts of the year. This relatively simple and potentially inexpensive method is to deposit spent fluids and cuttings directly on the ice. The method has recently been tested (Miller et al., 1982). As in ocean discharges, the fates of drilling fluids and cuttings disposed of on ice depend on site-specific conditions. Discharges deposited on ice in nearshore areas subject to overflow flooding from rivers would be widely dispersed during the annual breakup of the ice. Without such flooding, the discharges are dispersed more gradually. Liquid fractions are removed during initial surface melting. Depending on the movement of the ice, solids may be either deposited near the disposal site or carried with the ice and widely deposited over the seafloor.

## SUBSTITUTIONS

### Altered Composition

Just as special fluids are formulated for special downhole conditions, the composition of drilling fluids can be altered to include less toxic compounds for environmental reasons. For example, paraformaldehyde, a nonpersistent biocide, is used instead of chlorinated phenols on the OCS. Comparable substitutions have been developed for lubricants (for example, paraffinic oils for diesel fuel). In replacing diesel fuel, which is toxic, the use of other additives, such as emulsifiers, can also be minimized. (An important purpose of emulsifiers is to integrate diesel fuel with other components of the drilling fluids.)

A variety of mineral and vegetable oil-based products have been developed as alternatives to petroleum hydrocarbons as drilling fluid additives. While these products are in limited use in the U.S., they are used more extensively elsewhere. Field and laboratory tests of operating characteristics and environmental acceptability have been conducted. Tests and trial introductions continue as experience is gained concerning the alternatives' operating characteristics and environmental acceptability.

In substituting other oils for diesel fuel (and at other times), it may be desirable to monitor the composition of drilling fluid to quantify and distinguish between various types of hydrocarbons. Gas chromatographic methods, such as those developed by ASTM Committee

D-19 can be used. Methods given in part 31, ASTM Standards (1982) can possibly be applied to whole oils, waterborne oils, and marine sediments to examine the composition, quantities and origin of hydrocarbons.

### Processing Drilling Fluids Prior to their Discharge

Drilling fluids are processed while they are used to separate cuttings (see Chapter 2). Better use of solids control equipment can in some instances reduce the total volume discharged. So-called "closed mud systems", available commercially, accomplish this. Drilling fluids cannot be reused more extensively than they are in current practice because of the need to condition them for desired functions. It is always in the operator's economic interest to conserve and reuse drilling fluids when possible, but the operator must occasionally dispose of a fluid to use another with more appropriate characteristics for a given operating condition.

Each of these alternatives is characterized by different costs and risks than those of common practice. The costs of such alternative practices tend to be higher, although in some instances only slightly so.

A more serious concern about such alternatives is their risk. The environmental fates and effects of some alternative fluids and additives may be less well known than those of fluids commonly used. From an operational standpoint, the alternatives may require different operating techniques and handling than commonly used fluids. When using alternatives, operators cannot rely on the training and experience they have had with commonly used fluids.

Alternative additives and processes may offer advantages in special situations, but until more experience with them is acquired, their operating and environmental benefits and risks will not be established.

## OTHER ALTERNATIVES

The remaining alternatives (in Table 24), incineration and injection, are not considered practicable. Incineration is not because large amounts of the discharge are incombustible (OOC, 1981). Injection of other than the liquid fraction of drilling discharges into porous formations is not technically feasible, since one property of drilling fluid is to consolidate loose formations encountered while drilling, thus clogging the pores and preventing the formation from accepting new material. To inject under these conditions would require high pressures, and even then the formations would resist accepting the material.

## THE "NO DISCHARGE" ALTERNATIVE--A CASE STUDY

Alabama's regulations for offshore exploratory drilling in Mobile Bay prohibit the discharge of solid and liquid wastes like drilling fluids, drill cuttings, sand, contaminated deck drainage, and effluents from sewage treatment units. Only uncontaminated rainwater and water from the bay used to preload rig legs or to test fire-fighting equipment can be discharged. A major oil company recently accepted these conditions and proceeded with a drilling program using the "no discharge" alternative.[3] In the course of planning, alternatives and costs were considered in detail. Collection and disposal of waste materials while drilling in shallow Mobile Bay present unique and costly problems. An obvious solution is to collect all wastes in barges and transport them to shore for disposal. However, other methods appeared feasible to the company and were investigated.

Since the wastes this drilling generated might be discharged to the sea under an NPDES permit if the drilling occurred on the OCS, barging these wastes to federal waters for ocean disposal was one alternative considered. However, long lead times were anticipated in obtaining the required designation of an ocean dump site. The delay was sufficient to rule out ocean dumping for the initial well, but there were also questions to address about the ocean dumping equipment. Could barges used in other ocean dumping contain the liquids without seeping as these liquids were collected at the drill site? If not, could an adequate sealing system be developed? Could more appropriate barges be developed in the time available? These questions were never answered.

Since most drilling wastes would be liquid, another form of disposal considered was the subsurface injection of liquid wastes at the drilling site and solids disposal onshore. Alabama regulatory agencies indicated that a permit could be obtained for an onsite disposal well for well fluids and contaminated deck drainage, but that injection of sewage treatment effluents would require a permit that the state would not issue. Specifying an additional onsite injection well in the plans that were already under review by regulatory agencies would have delayed obtaining needed permits. Since this option only partly solved the problems of liquids disposal, the anticipated delay was unacceptable.

Disposal of all wastes, liquid and solid, at onshore disposal sites was carefully evaluated, with potential sites inspected by two company teams. In this as in similar cases a problem was posed by the limited number of acceptable active facilities. Some sites considered were only in the planning stage. Permits had not been obtained in some cases, and some operators had little or no experience in managing waste facilities. Some active sites were judged unsuitable because of operational practices.

---

[3] Floyd Garrot, Exxon, personal communication, January 1983.

The only active disposal facilities that met all company criteria were sites for hazardous wastes. The main disadvantage for the company of using these sites was a much higher disposal cost. A second disadvantage in using these facilities may be the use of limited disposal space that might be needed for hazardous materials.

The plan finally adopted by the company for its first well specifies disposal of solids by landfill and of liquids by subsurface injection. Both services are provided at a hazardous waste disposal facility located in Port Arthur, Texas, approximately 740 km from the drilling site. Waste solids are buried in clay-lined pits. Waste liquids, containing up to 10 percent suspended solids by volume and less than 200 ppm oil and grease, are first filtered to remove the suspended solids for burial. The filtered liquids are then injected into a disposal well 2,200 m deep. Other liquids require solidification and landfill burial. The wastes are transferred from barges to trucks at dock facilities about 30 km from the disposal facility. Backup disposal capability is available at a hazardous waste disposal landfill in Alabama approximately 390 km north of the drill site. While closer to the drill site, this facility does not have an injection well. Disposal costs would be much higher at this site because of the need to solidify all liquids for burial.

The jackup drilling rig used was specially designed and built for the "no discharge" operation. This increased the contractor's construction cost by about $635,000 to add features not normally required for drilling in the Gulf of Mexico. The cost of retrofitting a rig not specially designed and built for the operation would be a minimum of $1 million. The cost of bringing an active rig in for modification, including standby and transportation charges, could be as much as $3 to $4 million. Special rig features for this operation include extensive use of coaming, drip pans, and drains to capture and collect liquids from all equipment areas and drainage surfaces and manifolding the drain lines of the shale-shaker tank and the cuttings chutes. Piping to divide the solid and liquid wastes for separate barges was installed at the well site. Uncontaminated rainwater is kept separate from other liquids to be used in the drilling fluid. Temporary onboard storage space for liquid wastes was also provided to handle anticipated short periods when barges might not be available for immediate discharge of the wastes. Five tank barges, three hopper barges, and one tug are used full time to handle waste collection and transportation. When drilling a large-diameter hole, an additional tug is needed. The barges required modifications to prevent pollution. Coaming (raised framing for capturing and directing runoff or spills) was installed around the pump and discharge lines on the tank barges, and the hopper barges had to be compartmentalized to stabilize their cargoes. In addition, mooring anchors and piling were required to maintain the barges in position at the rig.

Waste drilling fluids, cement, contaminated drill-floor deck drainage, and formation cuttings are collected in an open-top, compartmentalized hopper barge. When filled, the receiving barge is replaced by another. The full barge is towed to Port Arthur, Texas. The wastes are transferred to trucks for transportation to the disposal site.

Contaminated rainwater and deck drainage (other than from the drill floor), effluents from the sewage treatment unit, wash water and other liquid wastes are collected at the well site in a permanently moored, compartmentalized tank barge. The liquids are periodically transferred to other tank barges for transportation to Port Arthur, and then trucked to the disposal site.

Table 26 gives the estimated additional costs of this special disposal operation.

TABLE 26 Estimated Costs of the "No Discharge" Alternative

| | Estimated Costs (Thousands of Dollars) | |
|---|---|---|
| | Minimum[a] | Maximum[b] |
| Rig modifications | - | - |
| Barge modifications | 500 | 500 |
| Tug and barge rental (including fuel) | 720 | 1,000 |
| Barge mooring (pilings and anchor systems) | 535 | 535 |
| Barge rig-up on location | 250 | 250 |
| Extra supervision and technical support | 250 | 360 |
| Waste facility charges for disposal | | |
| Liquids | 430 | 780 |
| Solids | 380 | 570 |
| Total | 810 | 1,350 |
| TOTALS[c] | 3,065 | 3,995 |

[a]Assuming: 250 drilling days and a disposal volume of 110,000 bbl (sewage treatment, 35 bbl; contaminated deck drainage and rainwater, 55 bbl; solids from drilling fluids and cuttings 30,000 bbl.

[b]Assuming 350 drilling days and a disposal volume of 190,000 bbl (sewage treatment, 35 bbl; contaminated deck drainage and rainwater, 110,000 bbl; solids from drilling fluids and cuttings, 45,000 bbl).

[c]The estimated total cost of the well is greater than $20 million.

## REFERENCES

Ayers, R.C., Jr., T.C. Sauer, Jr., D.O. Stuebner, and R.P. Meek. 1980. An environmental study to assess the effect of drilling fluids on water quality parameters during high rate, high volume discharges to the ocean. In: Proceedings of a Symposium on Research on Environmental Fate and Effects of Drilling Fluids and Cuttings. Washington, D.C.: Courtesy Associates. Pp. 351-381.

Grant, W.D., and O.S. Madsen. 1982. Moveable bed roughness in unsteady oscillatory flow. J. Geophys. Res. 87. Pp. 469-481.

Miller, R.C., R.C. Britch, R.V. Shafer, and S.O. Hillman. 1982. How offshore arctic conditions affect drilling mud disposal. Pet. Eng. Int.:68-88.

Offshore Operators Committee (OOC). 1981. Alternate disposal methods for muds and cuttings Gulf of Mexico and Georges Bank. Available from Hubert Clotworthy, Chairman, Environmental Subcommittee, Texaco Inc., P.O.B. 60252, New Orleans, LA 70160.

# Appendix A
# Composition of Drilling Fluids

## DENSITY MATERIALS

Materials used in drilling fluids in the greatest quantities are those added to increase the density in order to control subsurface pressures, which increase with depth as the well is drilled. The dense column of fluid exerts considerable pressure on the bottom of the borehole to keep formation fluids and gases from entering the borehole in an uncontrolled fashion. Several characteristics of these density materials are common among the various compounds and are essential for optimal performance. These materials should be (1) of high specific gravity, (2) nonreactive with the liquid phase of the fluid, (3) nonabrasive, and (4) of optimal particle size.

High specific gravity is required in order to maximize the weight or density in the smallest volume possible for logistical and economic reasons. Nonreactivity is essential because the weighting agent must be added oftentimes in increasing amounts as the well deepens and yet fluid properties (e.g., rheology or fluid behavior) must not be significantly affected by the influx of weighting materials. In the same sense, it is important that the weighting agent not be abrasive in pumps or to drill stages, particularly at higher concentrations, in order to reduce wear. Optimal particle size is essential for several reasons. This includes a minimal value ($>2\mu$) below which the particles affect mud properties easily (e.g., viscosity) and an upper value ($<44\mu$) beyond which the particles are hard to keep in suspension and also tend to become abrasive. By balancing these various needs, one can obtain an optimal product. Some products, such as lead or galena, are only used in special cases (e.g., to "kill" an uncontrolled well) while others are used primarily only in certain fluid systems (e.g., calcium carbonate in low density oil base or emulsion drilling fluids).

Of all of the materials, barite is by far the most commonly used weighting product worldwide. Indeed, it accounts for the largest proportion of all components which are in drilling fluids. Barite has fairly high specific gravity (4.3-4.5) but is also fairly soft and nonabrasive. It is inert in both oil and water.

The amount of barite used annually depends on the general drilling activity and particularly on drilling activity in high pressure zones.

In 1942, the domestic use of barite barely exceeded 100,000 tons per year, but it increased to 1,900,000 in 1978 (API, 1978b) and may well reach 3,000,000 tons by the year 2000 (Morse, 1981). The United States is the world's largest producer and consumer of barite. Mining of barite takes place in nine states with Nevada, Arkansas, and Missouri accounting for most of the production. Additional sources outside of the U.S.A. include Peru, Ireland, China, India, Mexico, Morocco, and Thailand.

Barite is barium sulfate ore. Most of the barite used in drilling fluids contains 80-90 percent $BaSO_4$. It is known as barytes, heavy spar, tiff, and cawk throughout the world and is surface or shaft mined mainly from vein, residual, or bedded deposits. The barite ranges in color from white to black and may be interspersed with a variety of other minerals (e.g., quartz, clay, pyrite) that may constitute from 10-15% of barite. The barite is separated from these materials at the mine, if necessary, by a series of devices which collectively enrich the amount of pure barite. The final product is dried and ground at the mine or at separate grinding plants throughout the world and packaged or sold in bulk.

Barite that is used in oil well drilling is required to conform to a set of specifications established by API (1981) in order to ensure consistent material. These specifications stipulate a minimum specific gravity, pore size range and alkalinity.

Other than barite, few other compounds are used to any extent in the domestic market as density materials. Iron oxide or hematite ($Fe_2O_3$) was one of the first materials used but its use has been largely discontinued because of its characteristic staining of skin and clothes of the drilling crews. It is currently seeing some revived interest as a product or as an additive to barite.

## VISCOSIFIERS

In order for the drilling fluid to remove the formation solids and cuttings from the bit at the bottom of the hole and carry them up the annulus to the surface, the fluid must have a certain thickness or viscosity. Only the smallest of particles could be carried up the long column with just pure water, even under pressure. More importantly, pressure or flow must be interrupted during the drilling process (for example, to change the drill bit) and pure water alone would allow the solids to fall back to the bottom during these static periods. It is essential, therefore, that the drilling fluid be viscous enough to suspend the cuttings during these periods..

As discussed in the "Drilling Discharges" chapter, drilling fluid ingredients often have several functions. Viscosifiers also help seal the wellbore and prevent loss of liquids to the formation, an essential function because an uncontrolled fluid loss would require constant monitoring and addition of water to correct for the loss. More importantly, critical fluid properties would be in a state of constant flux and would make it difficult to drill. It is difficult to separate ingredients in these two groups because of their functional overlap.

Clays, however, will be discussed under viscosifiers and polymers will be discussed as fluid loss agents event though both can functional dually.

The primary viscosifiers are the various clays that are added. Gray et al., (1980: 536) indicates that the term clay has several meanings; however, it is best defined to be those natural earth materials of fine grain size which are primarily composed of hydrous aluminum silicates. The definition becomes less precise as one reviews the clays; however, it is more important to understand their function and chemistry rather than find an exact definition.

In the mineralogical sense, Joseph (1978) indicates that drilling fluid clays fall into two groups--the smectite group (layered) and the hormite group (fibrous). The mineralogical nomenclature is confused by a number of regional and historical names. Since the most important characteristics of the clays are their bonding with other components, which is largely dictated by the polyvalent ion, the clays are often listed within each mineralogical group by cation where appropriate.

A variety of other classification schemes are available (Brindley, 1955; Degens, 1965; Warshaw and Roy, 1961), which use a compositional basis for classification, and may be referred to for more detailed information on clay structure. Since clays are natural minerals, the clay composition and the presence of impurities are variable due to geological differences. Silica, shale, calcite, mica, and feldspar are the most common impurities (Perricone, 1980).

The primary clay in drilling fluids and one of the earliest ones to be used is bentonite, composed mostly of sodium montmorillonite. It is still the most commonly used clay with domestic use in excess of 650,000 tons (API, 1978). Most of the domestically used bentonite is mined in Wyoming, South Dakota, and Montana. There mines are shallow, surface mines where the bedded clays are typically lenses between shale layers. The lenses are of variable thickness (e.g., 1-60 ft.) and under fairly shallow overburdens. The bentonite is mined, weathered, and then sized and dried before grinding. As with barite, the American Petroleum Institute maintains industrial specifications concerning manufacture and operating properties. These include moisture content, maximum particle size, and viscosity and plasticity criteria. Bentonite may be treated with a variety of polymers to produce viscosities equal to or greater than API specifications (Perricone, 1980).

Another bentonite, calcium montmorillonite (also called sub-bentonite) is used, but in far less quantities, due to its poorer performance than sodium montmorillonite. Often, sub-bentonite is added to bentonite when larger particle sizes are required. Other clays, such as attapulgite and sepeolite, are used as viscosifiers in salt water fluids. These clays are mined domestically also (Georgia, Florida, and Nevada, respectively). Due to their fibrous nature, these clays are not suitable for fluid loss reduction and serve also exclusively for viscosity.

## FLUID LOSS ADDITIVES

As discussed in the section on viscosifiers, clays and various polymers act as both viscosifiers and as fluid loss additives. Any dissociation of the two is spurious; however, this discussion separates the two purely for clarity. A properly designed drilling fluid should deposit a filter cake on the wall of the well bore during drilling to retard the continuous liquid phase in the drilling fluid from entering the formation. Bentonite and drilled clays are the prime builders of this cake, but in some instances, they are inadequate to stop the flow. The addition of fluid loss control additives may be necessary. Polymers are particularly well suited to this task and both natural and synthetic polymers are used. Starch was the first natural polymer used (Gray et al, 1942). Corn and potatoes are the principal source of starch for drilling fluids. Starch is prepared by treating the raw materials with heat and chemical agents to gelatinize the starch and then dried and ground for bagging. Modified starches have also been developed to include cyanoethylated starch, amino starch ether, hydroxypropyl starch ether, and quatenary ammonium salts of starch. Additionally, preserved or nonfermenting starches are available which may include a biocide such as paraformaldehyde.

Several natural gums have been used. Guar gum from the guar plant, a Texas legume, is one of the most prevalent. Semi-synthetic gums produced from the chemical modification of cellulose comprise another prevalent group. Sodium carboxymethylcellulose (CMC) is widely used as well as hydroxyethylcellulose (HEC). These compounds all function by adsorbing to the omnipresent clays. Cellullosic polymers can be further treated to produce polyanionic forms usually with specific molecular weight ranges that function in the presence of salt when CMC is less effective. Synthetic polymers that are water dispersible have been developed in recent years that function very well in drilling fluids. Polymerization of acrylic polymers and acrylates has resulted in the development of a number of different additives that counter fluid loss, or function as flocculants, viscosifiers, or bentonite extenders. The chemistry, concentration, temperature, and original source material of the polymer additives are all variable, as well as critical to operating characteristics.

## THINNERS AND DISPERSANTS

The next largest group of products after viscosifiers are those products which act to reduce viscosity. As drilling proceeds, the drilling fluid has a tendency to thicken naturally from the addition of very fine formation solids and native clays. This tendency increases with depth as temperature rises due to the geothermal gradient and causes the mud to undergo high temperature gellation or thickening. Other chemical reactions may also create thickening of the mud. A change in the clay's surface chemistry is usually the cause of gelling and creates the need for a material to disperse the clay particles. Thinners typically have a relatively large anionic component which is adsorbed on the positive sites of the clay particles, thereby reducing the attractive forces between the particles. See Gray et al, (1980:164) for a more complete discussion of the thinning mechanism. Materials commonly used as thinners in water-based muds are broadly classified as plant tannins, polyphosphates, lignitic materials, and lignosulfonates. Tannins occur in many plants and are extracted from bark, wood, or fruit. Most of the tannins used in drilling fluids are from the extract of quebracho wood, one of the first thinners ever used in the United States (Lawton et al, 1933, 1935). Chemically, tannins are esters of one or more polyphenolic acids.

Polyphosphates are those phosphates in which two or more phosphorus atoms are joined together by oxygen atoms, such as sodium tetraphosphate. Polyphosphates may be of varying chain lengths and the formula is usually expressed as the ratio of $Na_2O/P_2OS$. Phosphonic acids and pyrophosphates are two other general types occasionally used. Both the tannins and phosphate compounds have temperature limitations. The phosphates become nonfunctional above 250$^{o}$F and the tannins degrade between 250-350$^{o}$F (Carney and Harris, 1975).

Lignitic materials include a variety of materials which chemically differ due to source and preparation. Variously called lignite, leonardite, mined lignin, brown coal, and slack, these materials became popular as thinners after World War II when quebracho exports were diminshed. Lignite and brown coal are actually low heat value coals while leonardite is a naturally oxidized lignite from prolonged weathering. Leonardite has a high content of humic acid; several grades are available with varying humic acid contents. North Dakota is the principal source with South Dakota, Montana, New Mexico, and Texas as secondary sources. The material is usually strip mined, dried to 15-20% moisture content, crushed and bagged. Modified lignites were found to be excellent thinners after treatment with caustic soda, chrome or potassium salts to produce a more temperature stable compound. Lignitic thinners can perform satisfactorily at high temperatures (350$^{o}$F) and are often used in geothermal drilling fluids.

The lignosulfonates are waste byproducts of the sulfite process for pulping wood to make paper. The chemistry in making lignosulfonates and the chemical structure of lignosulfonates are complex but covered in a number of books and articles (Browning and Perriane, 1962; Carney, 1970; and Sarkanen and Ludwig, 1971). In simplest terms, lignosulfonates are polymeric salts of lignosulfonic acids with various functional compounds attached. The functionality of the lignosulfonates are enhanced by the functional compounds. These are usually added during the sulfite pulping process and reacted, then recovered by spray drying. Calcium, chrome, and iron compounds are the predominant materials added to form calcium chromium, ferrochromium, or ferrolignosulfonates. The mechanism of thinning by lignosulfonates is discussed in Jessen and Johnson (1963) and more completely in Gray et al (1980: Chapter 4), but also relies on the lignosulfonate micell attaching to the edge surface of the clay particles to break the electrokinetic attraction between the clays. As the clays disperse, the viscosity of the fluid decreases and the fluid is "thinned".

## pH AND ION CONTROL

The pH of most water based drilling fluids is kept alkaline for a variety of reasons. Corrosion control and the control of poisonous $H_2S$ gas are two of the prime reasons; however, other fluid properties are also affected (e.g., solubility of additives). For these reasons, an alkaline pH is maintained by adding caustic soda (sodium hydroxide) to the system as needed. Ionic balance is also commonly affected by contamination of the fluid system by cement, salt, or anhydrite. These inputs can affect the rheology (fluid behavior) of the system and require treatment. Soda ash (sodium carbonate) and baking soda (sodium bicarbonate) are the most common additives. A number of additional materials may be used for specialized fluids or very specialized problems, but their relative usage frequency is rather small in comparison to the above three materials: barium carbonate, potassium hydroxide, calcium sulfate, calcium hydroxide, sodium chloride, and potassium chloride.

## LUBRICANTS

Under normal drilling, the drilling fluid alone is sufficient for adequate lubrication of the drill pipe and bit. However, because no hole is truly vertical and the drill pipe is flexible, there are likely to be some points of contact between the side of the hole and the drill pipe. This creates a frictional resistance thereby increasing the torque required to turn (as well as raise and lower) the drill pipe and bit. Lubricants are added to drilling fluids when friction is encountered. The addition of lubricants is generally required for highly deviated holes, holes with frequent direction changes, undergauge holes, or holes with poor drill string dynamics.

Oil base drilling fluids are excellent lubricants, however, due to higher costs and government regulations, oil fluids are generally not used where the only advantage is lubrication. A common historical practice concerning lubricants has been to add diesel fuel (No. 2 fuel oil) to a water base fluid. Fluids with high diesel fuel content (as much as 50 percent) may be used to counter friction. Prior to discharge, these fluids will be worked to separate the diesel fuel from the discharge, or the fluid may be diluted to lower the relative diesel fuel content. These practices are within existing regulations, so long as the drilling discharges do not cause a sheen on the surface of the ocean or a sludge on the seafloor. Moreover, with detergents and/or emulsifiers in the system, fluids containing as much as several percent diesel can sometimes be discharged without a sheen. More importantly, laboratory testing has demonstrated that emulsified oil in a water based fluid is not as effective a lubricant as non diesel substitutes currently available. Also, these substitutes may have less effect on the fluids' rheology than does emulsified oil. A list of chemicals used as lubricants is in Table A-1.

## LOST CIRCULATION MATERIALS

Circulating drilling fluids can be lost to downhole formations through induced fractures, preexisting open fractures, caverns, pores, and solution channels. Lost circulation is one of the most common problems encountered during rotary drilling. Lost circulation materials are added to the mix either as an additive or in some cases as a premixed slurry slug. Whatever the means employed to add the material to the fluid, the end result is the same--that of actually plugging the fractures or openings. These additives are either fibrous, filamentous, granular, or flaked and are almost always naturally occurring. Common lost circulation materials include ground nut shells, mica, and ground cellophane.

## CORROSION INHIBITORS

Corrosion of downhole tubular pipe is a very serious problem. The simplest and most common means to control corrosion is to use a highly alkaline drilling fluid, but this practice has limitations--hydroxyl ions degrade clay minerals at temperatures above 200°F and a pH above 10. There are three major forms of corrosion:

- Carbon Dioxide. $CO_2$ dissolves in water resulting in a lowering of pH values through the production of carbonic acid. This can be controlled using sodium hydroxide to a pH of 9-10. In some cases excessive acids may be produced; these can be neutralized with calcium hydroxide, but this can precipitate scale deposits which set up corrosion cells. Scales can be controlled by the addition of a scale inhibitor such as sodium phosphonate.
- Oxygen. $O_2$ is almost always present in drilling fluids where only a minimal amount is sufficient to cause significant corrosion pitting under rust or scale patches. This form of corrosion is controlled using oxygen scavengers such as sodium sulfite or ammonium bisulfite. Filming amines and morpholines can also mitigate corrosion by deposition of a film on metal surfaces. Chromates can also be used to incorporate a film (a complex of oxygen, iron, and chromium) on downhole metal surfaces. Hexavalent chrome is reduced to trivalent chromium in the fluid system.
- Hydrogen Sulfite. $H_2S$ may contaminate the drilling fluid by an influx of sour gas or by the degradation of lignosulfonates by sulfate-reducing bacteria or by high temperatures (330°F). Hydrogen sulfide is both a deadly poison and a severe source of corrosion, through the mechanism of hydrogen embrittlement. It can be removed from the drilling fluid system through the use of sulfide scavengers such as zinc carbonate, zinc oxide, or organically chelated zinc compounds which prevent mud flocculation by reducing the amount of free zinc ions. Iron oxides are also used and do not affect the rheological or filtration properties of the mud.

TABLE A-1 Chemicals Commonly Used in Drilling Fluid Lubricants

| | |
|---|---|
| Acetophenones | Lanolin |
| Alcohol Ester | Low Paraffinic Solvents |
| Aluminum Stearate | Mineral Oil |
| Asphalts | Organic Phosphate Ester |
| Calcium Oleate | Rosin Soap |
| Coconut Diethanolamides | Sodium Alkylsulfates |
| Coconut Oil Alkanolamide | Sodium Asphalt Sulfonate |
| Diesel Fuel | Sodium Phosphates |
| Diphenyl Oxide Sulfonate | Sorbitan Ester Sulfonate |
| Ethoxylates | Stearates |
| Ethoxylated Alcohol | Sulfonated Alcohol Ether |
| Fatty Acid Soaps | Sulfonated Tall Oil |
| Gilsonite | Sulfonated Vegetable |
| Glycerol Dioleate | Triethanolamine |
| Glycerol Monoleate | Vegetable Oils |
| Glass Beads | Wool Greases |
| Graphite | |

## BACTERICIDES

Three mechanisms can be used to prevent or mitigate fermentation of drilling muds by microorganisms. Saturated salt muds and highly alkaline muds (ph $\geq$ 12) are resistant to bacteriological activity. If these options are not available, then the addition of a bactericide may be necessary. Bactericides are most common in drilling fluids containing starch or polymers which are rapidly degraded by heat, agitation, or microorganisms. Paraformaldehyde is the most common bactericide used in drilling fluids as well as workover and completion fluids.

Paraformaldehyde will depolymerize in acidic or basic solutions to form its monomer, formaldehyde. Neglecting any loss due to absorption, there are several means by which it will be depleted from the drilling fluid system. Reactions with bacterial mucoproteins with activated aromatic rings found in lignosulfonates and tannins reduce the paraformaldehyde. Also, destruction by air oxidation forms either a formate or $CO_2$ by the Cannizaro reaction. Due to these mechanisms, paraformaldehyde must be routinely added to the system to maintain adequate treatment levels. Several non-fermenting starch additives are in use today. In these additives the bactericide has been incorporated into the starch, eliminating the need to add additional bactericide.

Under the current regulatory scheme, all bactericides used in drilling fluids are regulated by the Environmental Protection Agency under the Federal Insecticide, Fungicide, and Rodenticide Act (FIFRA) as well as discharge permits. Also, the Minerals Management Service has banned chlorinated phenols from use on the OCS. Under FIFRA registration, end-use, labeling, chemical identification, and application rates are tightly regulated. Of the many available bactericides, relatively few bactericides have oilfield registrations. Those without proper registrations are not used. Robichaux (1975) and Jones et al., (1980) listed eight different chemical groups that have been used (quartenary amines, paraformaldehyde, cupric sulfate, chlorinated thiophene chloride, glutareldehyde, carbonates, triaza chlorides, and chlorine dioxide). Additional materials (e.g., isothiazoline) have received approval since that time.

## SURFACTANTS

Surface active agents are adsorbed on surfaces and at interfaces resulting in a decreased surface tension. These are used in drilling fluids for several different purposes such as emulsifiers, wetting agents, foamers, defoamers, and agents to decrease the hydration of clay particle surfaces.

There are three main forms of surfactants. Cationic surfactants dissociate into large organic cations and simple inorganic anions. These are usually salts of a fatty amine or polyamine such as trimethyl dodecyl ammonium chloride. Anionic surfactants dissociate into large organic anions and simple inorganic cations. Soaps are the most common form such as sodium oleate. Nonionic surfactants are long chain polymers and do not dissociate. The most common nonionic surfactant is phenol reacted with 30-mol ethylene oxide. Cationic surfactants are strongly adsorbed on to negatively charged clay and rock surfaces, whereas anionic surfactants are adsorbed at the positively charged ends of clay crystal lattices resulting in a retardation of the hydration of bentonite. Other chemicals used as drilling fluid surfactants are found in Table A-2.

TABLE A-2 Chemicals Commonly Used in Drilling Fluid Surfactants

| |
|---|
| Coconut Diethanolamides |
| Coconut Oil Alkanolamide |
| Calcium Oleate |
| Fatty Acid Derivatives |
| Polycyclic Alcohol Anhydrides |
| Polyoxyethylenes |

## Emulsifiers

The relationship between emulsifiers and surfactants is direct. Interfacial tension between oil and water is very high but can be lowered through the use of a surfactant, which decreases the surface tension resulting in an emulsion--a stable dispersion of fine droplets of one liquid into another liquid. In addition, emulsifiers stabilize emulsions due to their molecules adsorbing at the oil/water interfaces forming a protective "skin" around dispersed droplets which prevents coalescing when these droplets collide.

Emulsifiers are generally used in oil-base fluids, however, they can be used in water-based systems to emulsify oil into the water phase. They can either act to emulsify oil into water or water into oil. Stable mechanical emulsions can be formed without using a chemical emulsifier (surfactants) by the adsorption of colloidal solids in the fluid at the oil-water interfaces. Dispersed clays and lignosulfontates can act as mechanical emulsifiers in alkaline fluids. Representative chemicals used as emulsifiers are found in Table A-3, as well as those listed under surfactants, Table A-2.

TABLE A-3 Chemicals Commonly Used in Drilling Fluid Emulsifiers

| |
|---|
| Alkyl Aryl sulfonates and sulfates |
| Polyoxyethylene fatty acids, esters, and ethers |
| Nonylphenol reacted with ethylene oxide |
| Fatty acid soaps, polyamines, and amides blends |
| Calcium Dodecylbenzene Sulfonates |

## FOAMERS AND DEFOAMERS

Foaming agents (surfactants) are added to remove borehole water while air drilling, to create a low-density fluid to remove drill solids when performing workover and/or completions in depleted reservoirs, or as an insulating medium in arctic wells. Very few wells drilled on the OCS use air or foam drilling due to the depth and pressures. Chemicals normally used are found listed under surfactants, Table A-2.

Defoamers are used to break foams used in drilling or those formed in gas-cut drill fluids. By far the most common chemicals used as defoamers are 2-ethyl hexanol, aluminum stearate, and ester alcohols such as the monoisobutyrates.

## FLOCCULANTS

Flocculants are used to remove small cuttings when clear water drilling of hard rock is required. These flocculants can be injected in the fluid return after the shale shaker allowing solids to flocculate in the reserve pit. Flocculants are also used to clarify reserve fluid pits prior to disposal. They are rarely used offshore.

Acrylic polymers are excellent flocculants at a concentration of 0.01 lbs/bbl, but can perform a dual function as a filtration control agent at concentrations of 3 lbs/bbl. Chemicals currently used as flocculants include alum, calcium sulfate, polyacrylamides, sodium polyacrylate, copolymers of vinyl acetate and maleic anhydride, and calcium oxide.

## REFERENCES

American Petroleum Institute. 1978. Oil and gas well drilling fluid chemicals. Bulletin 13F, first ed. American Petroleum Institute, Washington, D.C. 9 pp.

American Petroleum Institute. 1981. API specifications for oil-well drilling-fluid materials. Bulletin 13A, eighth ed. American Petroleum Institute, Washington, D.C. 14 pp.

Brindley, G.W. 1955. Identification of clay minerals by x-ray diffraction analyses. Clays and Clay Technol. Bull. 169:119-129.

Browning, W.C., and A.C. Perricone. 1962. Lignosulfonate drilling mud conditioning agents. SPE Paper 432, Full Meeting, Society of Petroleum Engineers, October 7-10, 1962, Los Angeles, Calif. 13 pp.

Carney, L.L. 1970. Preparation and use of chromium lignosulfonates in drilling mud. Presentation to Spring Meeting, Production Department, American Petroleum Institute, March 4-6, 1970, Dallas, Tex. 9 pp.

Carney, L.L., and L. Harris. 1975. Thermal degradation of drilling mud additives. In: Proceedings, Environmental Aspects of Chemical Use in Well-Drilling Operations, May 21-23, 1975, Houston, Tex. EPA-560/1-75-004. Office of Toxic Substances, U.S. Environmental Protection Agency, Washington, D.C. vi + 604 pp.

Degens, E.T. 1965. Geochemistry of Sediments. Englewood Cliffs, N.J.: Prentice-Hall. 342 pp.

Gray, G.R., J.L. Foster, and T.S. Chapman. 1942. Control of filtration characteristics of salt water muds. Trans. AIME 146:117-125.

IMCO Services. 1980. Know your drilling mud components. Drilling Contractor 36(3):92-110.

Jessen, F.W., and C.A. Johnson. 1963. The mechanism of adsorption of lignosulfonates or clay suspensions. Soc. Pet. Engi. J. 3(3):267-273.

Jones, M., C. Collins, and D. Havis. 1980. Trade-offs in traditional criteria vs. environmental acceptability in product development: an example of the drilling fluids industry's response to environmental regulations. The Environmental Professional. Vol. 2. Pp. 94-102.

Joseph, J.H. 1978. Bentonite, sepiolite, and attapulgite. In: Raw Materials for the Oil Well Drilling Industry. P.W. Harben (ed.), Metal Bulletin LTD. London, England. Pp. 51-74.

Lawton, H.C., H.A. Ambrose, and A.G. Loomis. 1932. Chemical treatment of rotary drilling fluids. In: Physics. Pp. 365-375.

Lawton, H. C., A. G. Loomis, and H. A. Ambrose. 1935. Drilling wells and drilling fluid. U.S. Patent No. 1, 999, 766, April 30, 1935. U.S. Patent Office, Washington, D.C.

Morse, D. E. 1981. Barite. In: Mineral Facts and Problems. Bulletin 671. Bureau of Mines, U.S. Department of the Interior. Washington, D.C. 12 pp.

Perricone, C. 1980. Major drilling fluids additives. In: Proceedings of a Symposium on Research on Environmental Fate and Effect of Drilling Fluids and Cuttings. Courtesy Associates: Washington, D.C. Pp. 15-29.

Robichaux, T.J. 1975. Bactericides used in drilling and completion operations. In: Proceedings, Environmental Aspects of Chemical Use in Well-Drilling Operations, May 21-23, 1975, Houston, Tex. EPA-560/1-775-004. Washington, D.C.: Office of Toxic Substances, U.S. Environmental Protection Agency. vi + 604 pp.

Sarkanen, K.V., and C.H. Ludwig. 1971. Lignins: Occurrence, Formation, Structure and Reactions. New York: Wiley-Interscience. 451 pp.

Warsaw, C.M., and R. Roy. 1961. Classification and scheme for the identification of layer silicates. Geol. Soc. Am. Bull. 72:1455-1492.

# Appendix B
# Functionally Equivalent Drilling Fluid Products

The extent of redundancy of drilling fluid products is difficult to measure; nevertheless it would be useful information, for reviewing well histories for the purpose of permit compliance, for example. The drilling fluid products offered by each of the four major drilling fluid supplier companies is listed in Table B-1. While this list is for only a segment of the industry, it is a useful comparison of product equivalency because most of the offshore wells, and especially wells on the OCS, are serviced by the major supplier companies. Some areas (e.g., Georges Bank) have been exclusively serviced by majors. Of the 86 different drilling fluid components listed, 66% were available from all 4 companies, 7% from 3 of 4, 6% from 2 of 4, and only 6% were totally unique to one company. Products which are listed as functional equivalents may be different chemically, although a quick review of the table suggests that this is not often the case. It should also be noted that chemically equivalent products may be slightly different due to varying percentage of active ingredients, particle sizes, processing technique, relative proportion of components, or quality of materials. Nevertheless, this analysis suggests a rather substantial redundancy factor among the four largest companies. The inclusion in the table of the smaller companies which may not devote as much effort as the majors to product development, would likely not change the overall picture of redundancy provided by the table.

TABLE B-1 Comparable Drilling Fluid Products by Tradenames

| DESCRIPTION OR PRINCIPAL COMPONENT | IMCO SERVICES | BAROID | MAGCOBAR | MILCHEM | PRIMARY APPLICATION |
|---|---|---|---|---|---|
| WEIGHTING AGENTS - VISCOSIFIERS | | | | | |
| Barite | IMCO BAR™ | Baroid | Magcobar | Mil-Bar | For increasing mud weight up to 20 ppg |
| Barite/Hematite Blend | IMCO BAR-PLUS | Bar-Gain | | | For increasing mud weight up to 22 ppg |
| Hematite | IMCO NU-DENSE | | | | To increase density of a drilling and kill fluid up to 25 ppg |
| Calcium Carbonate | IMCO WATE | Baracarb | Lo-Wate | W.O. 35<br>W.O. 50 | For increasing density to 11 ppg with acid soluble material |
| Bentonite | IMCO GEL | Aquagel | Magcogel | Milgel | Viscosity and filtration control in water-base muds |
| Sub-Bentonite | IMCO KLAY | Baroco | High Yield Blended Clay | Green Band Clay | For viscosity and filtration control in water-base muds |
| Attapulgite | IMCO BRINEGEL | Zeogel | Salt Gel | Salt Water Gel | Viscosifier in saltwater muds |
| Beneficiated Bentonite | IMCO HYB | Quick-Gel | Kiwk-Thik | Super-Col | Quick viscosifier for freshwater, upper-hole muds with minimum chemical treatment |
| Asbestos Fibers | IMCO SHURLIFT | Flosal | Visquick | Flosal | Viscosifier for fresh-water or salt-water muds |

| | | | | | |
|---|---|---|---|---|---|
| Bacterially Produced Polymer | IMCO XC | XC Polymer | Duovis | XC Polymer | Viscosifier and fluid loss control additive for low-solids muds |
| Sepiolite | IMCO DUROGEL | | Geo-Gel | | Viscosifier in all water-base muds, especially high temperature drilling fluids |
| Multipurpose Polymer | IMCO POLYSAFE™ | | | Mil-Polymer 305 | Polymer for fluid loss control and viscosity |
| DISPERSANTS | | | | | |
| Sodium Tetraphosphate | IMCO PHOS (STP) | Barofos | Magco-Phos | Oil Fos | Thinner for low pH fresh-water muds where temperatures do not exceed 180°F |
| Sodium Acid Pyrophosphate | IMCO SAPP | SAPP | SAPP | SAPP | For treating cement contamination |
| Quebracho Compound | IMCO Q-B-T | Tannex | M-C Quebracho | Mil-Quebracho | Thinner for fresh-water and lime muds |
| Modified Tannin | DESCO | Desco | Desco | Desco | Thinner for fresh-water and salt-water muds alkalized for pH control |
| Processed Lignite | IMCO LIG | Carbonox | Tann A Thin | Ligco | Dispersant, emulsifier and supplementary additive for fluid loss control |
| Causticized Lignite | IMCO THIN | CC-16 | Caustilig | Ligcon | Dispersant, emulsifier and supplementary additive for fluid loss control |

TABLE B-1 (continued)

| DESCRIPTION OR PRINCIPAL COMPONENT | IMCO SERVICES | BAROID | MAGCOBAR | MILCHEM | PRIMARY APPLICATION |
|---|---|---|---|---|---|
| Chrome Lignosulfonate | IMCO-VC-10 | Q-Broxin | Spersene | Uni-Cal | Dispersant and fluid loss control additive for water-base muds |
| Blended Lignosulfonate Compound | IMCO RD-111 | | | | Blended multi-purpose dispersant, fluid loss agent and inhibitor for IMCO RD-111 mud systems |
| Chrome-Free Lignosulfonate | IMCO RD-2000™ | | Magco CFL | X-KB Thin | Dispersant and fluid loss control additive for water-base muds |
| FLUID LOSS REDUCERS | | | | | |
| Organic Polymer | IMCO PERMALOID | DEXTRID | Magco Poly Sal | | Control fluid loss in water-base muds |
| Pregelatinized Starch | IMCO LOID | Impermex | My-Lo-Gel | Milstarch | Controls fluid loss in saturated saltwater and lime muds |
| Sodium Carboxy-methylcellulose | IMCO CMC (Regular) | Cellex (Regular) | Magco CMC (Regular) | Milchem CMC (Med-Vis) | For fluid loss control and barite suspension in water-base muds |
| Sodium Carboxy-methylcellulose | IMCO CMC (High Vis) | Cellex (High Vis) | Magco CMC (High Vis) | Milchem CMC (High Vis) | For fluid loss control and viscosity building in low-solids muds |
| Polyanionic Cellulosic Polymer | DRISPAC | Drispac | Drispac | Drispac | Fluid loss control additive and viscosifier in salt muds |
| Polyanionic Cellulosic Polymer | DRISPAC SUPERLO | Drispac Superlo | Drispac Superlo | Drispac Superlo | Primary fluid loss additive, secondary viscosifier in water-based muds |

| | | | | | |
|---|---|---|---|---|---|
| Sodium Polyacrylate | IMCO SP-101 | Cypan WL-100 | Cypan WL-100 | Cypan WL-100 | Fluid loss control in calcium free muds |
| LUBRICANTS - DETERGENTS - EMULSIFIERS | | | | | |
| Specially prepared blend of organic liquid compounds | IMCO LUBE-106™ | | | | A water dispersable, non-foaming, nontoxic additive designed to impart lubricity and reduce torque, drag and friction in all water-base drilling fluids |
| Blend of Organic Esters | IMCO LUBRIKLEEN | Torq Trim II | DOS-3 | Mil-Plate 2 | Supplies the lubricating properties of oils without their environmental pollution |
| Extreme Pressure Lubricant | IMCO EP LUBE | EP Mudlube | Lube Bit Lube | Lubri-Film | Used in water-base muds to impart extreme pressure lubricity |
| Oil Soluble Surfactants | IMCO FREEPIPE | Skot-Free | Pipe Lax | Petrocote | Nonweighted fluid for spotting to free differentially stuck pipe |
| Blend of Fatty Acids, Sulfonates and Asphaltic Materials | IMCO SPOT™ | SF-100 | | Carbo-Free | Invert emulsion that may be weighted to desired density for placement to free differentially stuck pipe |
| Water Dispersible Asphalts | IMCO HOLECOAT II | | STABIL-HOLE | ITI-WD | Lubricant and fluid loss reducer for water-base muds that contain no diesel or crude oil |

TABLE B-1 (continued)

| DESCRIPTION OR PRINCIPAL COMPONENT | IMCO SERVICES | BAROID | MAGCOBAR | MILCHEM | PRIMARY APPLICATION |
|---|---|---|---|---|---|
| Processed Hydrocarbons | SOLTEX | Soltex | Soltex | Soltex | Used in water-base muds to lower downhole fluid loss and minimize heaving shale |
| Oil Dispersible Asphalts | IMCO MUD OIL | Baroid Asphalt | Pave-A Hole | Carbo-Seal | Lubricant and fluid-loss reducer for water-base muds that contain diesel or crude oil |
| Detergent | IMCO MD | Con Det | D-D | Milchem MD | Used in water-base muds to aid in dropping sand. Emulsifies oil, reduces torque and minimizes bit balling |
| Blend of Anionic Surfactants | IMCO SWS | Trimulso | Salinex | Atiosol and Atiosol S | Emulsifier for salt-water and fresh-water muds |
| DEFOAMERS - FLOCCULANTS - BACTERICIDES | | | | | |
| Aluminum Stearate | Aluminum Stearate | Aluminum Stearate | Aluminum Stearate | Aluminum Stearate | Defoamer for lignosulfonate muds |
| Liquid Surface Active Agent | IMCO DEFOAM-L™ | | | | Defoamer for all water-base muds |
| Surface-Active Dispersible Liquid Defoamer | IMCO FOAMBAN | Bara-Defoam 1 W300 | Magconol | LD-7 LD-8 | All-purpose defoamer |
| Flocculating Agent | IMCO FLOC™ | Barafloc | Floxit | Separan | Used to drop drilled solids where clear water is desirable for a drilling fluid |

| | | | | | |
|---|---|---|---|---|---|
| Blended Carbonate Solutions | IMCO CIDE | Bara-B33 | Magco Poly Defoamer | | Bactericide used to prevent fermentation |
| Paraformaldehyde | Para-formaldehyde | Aldacide | Paraformal-dehyde | Paraformal-dehyde | Bactericide used to prevent fermentation |
| LOST CIRCULATION MATERIALS | | | | | |
| Fibrous Material | IMCO FYBER | Fibertex | Mud Fiber | Mil-Fiber | Filler as well as matting material to restore lost circulation |
| Nut Shells: Fine | IMCO PLUG | Wall-Nut | Nut-Plug | Mil-Plug | Most often used to prevent lost circulation |
| Nut Shells: Medium | IMCO PLUG | Wall-Nut | Nut-Plug | Mil-Plug | Used in conjunction with fibers or flakes to regain lost circulation |
| Nut Shells: Coarse | IMCO PLUG | Wall-Nut | Nut-Plug | Mil-Plug | Used where large crevices or fractures are encountered |
| Ground Mica: Fine | IMCO MYCA | Micatex | Magco-Mica | Milmica | Used for prevention of lost circulation |
| Ground Mica: Coarse | IMCO MYCA | Milcatex | Magco-Mica | Milmica | Used for prevention and regaining of lost circulation |
| Cellophane | IMCO FLAKES | Jel Flake | Cell-O-Seal | Milflake | Used to regain lost circulation |

TABLE B-1 (continued)

| DESCRIPTION OR PRINCIPAL COMPONENT | IMCO SERVICES | BAROID | MAGCOBAR | MILCHEM | PRIMARY APPLICATION |
|---|---|---|---|---|---|
| Combination of granules, flakes and fibrous materials of various sizes in one sack | KWIK SEAL | Kwik Seal | Kwik Seal | Kwik Seal | Used where severe lost circulation is encountered |
| High-water loss slurry for lost circulation | Diaseal M | Diaseal M | Diaseal M | Diaseal M | Forms a high-solids plug to cure severe lost circulation |
| SPECIALTY PRODUCTS | | | | | |
| Bentonite Extender | IMCO GELEX | Benex | Benex | Benex | Increases yield of bentonite to form low-solids drilling fluid |
| Inhibiting Agent | IMCO IE PAC | K-Plus | | | Imparts inhibition, fluid loss and rheology control in potassium muds |
| Synergistic Polymer Blend | IMCO POLY Rx | Durenex | Resinex | | High-temperature rheological stabilization and filtration control |
| Biodegradable Surfactant | IMCO FOAMANT™ | Quick Foam | Magco Foamer 76 | Gel-Air | Foaming agent in air or mist drilling |
| CORROSION INHIBITORS | | | | | |
| Zinc Compound | IMCO SULF-X II | | | Mil-Gard | For use as a hydrogen sulfide scavenger in water-base and oil-base muds |
| Liquid Corrosion Inhibitor | IMCO CRACK-CHEK | | | | Prevent stress cracking of drill strings in an $H_2S$ environment |

| | | | | | |
|---|---|---|---|---|---|
| A Catalyzed Ammonium Bisulfite | IMCO $XO_2$ ™ | Coat 777 | OS-1L | Noxygen | For use an an oxygen scavenger |
| Filming Amine | IMCO X-CORR™ | Bara Cora | Magco Inhibitor | Aqua-Tec | All-purpose corrosion inhibitor |
| Filming Amine | IMCO PERMAFILM™ | Coat 415 Inhibitor | Magco Inhibitor | Ami-Tec | Corrosion inhibitor |
| Organic Polymer | IMCO SCALECHEK | Surflo-H35 | SL-1000 | Scale-Ban | Scale inhibitor |
| COMMERCIAL CHEMICALS | | | | | |
| Sodium Hydroxide | Caustic Soda | Caustic Soda | Caustic Soda | Caustic Soda | For pH control in water-base muds |
| Potassium Hydroxide | Caustic Potash | Potassium Hydroxide | Potassium Hydroxide | Potassium Hydroxide | Used to control pH in potassium system |
| Sodium Carbonate | Soda Ash | Soda Ash | Soda Ash | Soda Ash | For treating-out calcium in low pH muds |
| Sodium Bicarbonate | Sodium Bicarbonate | Sodium Bicarbonate | Sodium Bicarbonate | Sodium Bicarboñate | For treating-out calcium or cement in high pH muds |
| Barium Carbonate | Barium Carbonate | Anhydrox | Barium Carbonate | Barium Carbonate | For treating-out calcium sulfate (pH should be above 10 for best results) |
| Sodium Chromate | Sodium Chromate | Sodium Chromate | Sodium Chromate | Sodium Chromate | Used in water-base muds to prevent high-temperature gelation |

TABLE B-1 (continued)

| DESCRIPTION OR PRINCIPAL COMPONENT | IMCO SERVICES | BAROID | MAGCOBAR | MILCHEM | PRIMARY APPLICATION |
|---|---|---|---|---|---|
| Chrome Alum (chromic chloride) | Chrome Alum | Chrome Alum | Chrome Alum | Chrome Alum | For use in cross-linking XC Polymer systems |
| Calcium Sulfate | Gypsum | Gypsum | Gypsum | Gypsum | Source of calcium for formulating gyp muds |
| Calcium Hydroxide | Lime | Lime | Lime | Lime | Source of calcium for formulating lime muds |
| Sodium Chloride | Salt | Salt | Salt | Salt | For saturated salt muds and resistivity control |
| Calcium Chloride | Calcium Chloride | Calcium Chloride | Magcobrine C.C. | Calcium Chloride | For weighting solids-free brines and to control salinity in invert oil muds |
| Potassium Chloride | Potassium Chloride | Potassium Chloride | Magcobrine P.C. | Potassium Chloride | Potassium salt use in KCl inhibitive systems |